Getting Started with
Maple®
Third Edition

Douglas B. Meade
Michael May, S.J.
C-K. Cheung
G. E. Keough

*University of South Carolina,
St. Louis University and
Boston College*

EDITOR	David Dietz
MARKETING MANAGER	Sarah Davis
PRODUCTION MANAGER	Dorothy Sinclair
PRODUCTION EDITOR	Nicole Repasky
EDITORIAL ASSISTANT	Pamela Lashbrook
DESIGNER	Kevin Murphy

This book was set by Laserwords and printed and bound by Hamilton Printing.

Founded in 1807, John Wiley & Sons, Inc. has been a valued source of knowledge and understanding for more than 200 years, helping people around the world meet their needs and fulfill their aspirations. Our company is built on a foundation of principles that include responsibility to the communities we serve and where we live and work. In 2008, we launched a Corporate Citizenship Initiative, a global effort to address the environmental, social, economic, and ethical challenges we face in our business. Among the issues we are addressing are carbon impact, paper specifications and procurement, ethical conduct within our business and among our vendors, and community and charitable support. For more information, please visit our website: www.wiley.com/go/citizenship.

ISBN: 9780470455548

In loving memory of Kenneth D. Meade, my Dad

D.B.M

For my Parents

M.M., S.J.

To the memory of my Grandmoms

C-K.C.

For Dan and Ethel and Dan

G.E.K.

Preface

Using this Guide

The purpose of this guide is to give a quick introduction on how to use Maple. It primarily covers Maple 12, although most of the guide will work with earlier versions of Maple. Also, throughout this guide, we will be suggesting tips and diagnosing common problems that users are likely to encounter. This should make the learning process smoother.

This guide is designed as a self-study tutorial to learn Maple. Our emphasis is on getting you quickly "up to speed." This guide can also be used as a supplement (or reference) for students taking a mathematics (or science) course that requires use of Maple, such as Calculus, Multivariable Calculus, Advanced Calculus, Linear Algebra, Discrete Mathematics, Modeling, or Statistics.

About Maple

Maple is computer algebra software developed by Maplesoft, a division of Waterloo Maple Inc. This software lets you use the computer like an interactive mathematics scratchpad. Maple can perform symbolic computation as well as numerical computation, graphics, programming and so on. It is a useful tool not only for an undergraduate mathematics or science major, but also for graduate students, faculty, and researchers. The program is widely used as well by engineers, physicists, economists, transportation officials, and architects.

Organization of the Guide

The Guide is organized as follows:

- Chapter 1 gives a short demonstration of what you'll see in the remaining parts of the Guide. Chapters 2 through 11 contain the basic information that almost every user of Maple should know.

- Chapters 12 through 15 demonstrate Maple's capabilities for single-variable calculus. This includes working with derivatives, integrals, series and differential equations.

- Chapters 16 through 21 cover topics of multivariable calculus. Here you'll find the discussion focusing on partial derivatives, multiple integrals, vectors, vector fields, and line and surface integrals.

- Chapters 22 and 23 introduce the statistical capabilities of Maple.

- Chapters 24 through 27 address a collection of topics ranging from animation and simulation to programming, and list processing.

- Two appendices explain how to learn more about Calculus and a quick reference guide.

Chapter Structure

Each chapter of the Guide has been structured around an area of undergraduate mathematics. Each moves quickly to define relevant commands, address their syntax, and provide basic examples.

Every chapter ends with as many as three special sections that can be passed over during your first reading. However, these sections will provide valuable support when you start asking questions and looking for more detail. These three sections are:

- **More Examples**
  Here, you'll find more technical examples or items that address more mathematical points.

- **Useful Tips**
  This section contains some simple pointers that all Maple users eventually learn. We've drawn them from our experiences in teaching undergraduates how to use Maple.

- **Troubleshooting Q & A**
  Here we present a question and answer dialogue on common problems and error messages. You'll also be able to find out *how* certain commands work or *what they assume* you know in using them. Many useful topics are covered here.

Conventions Used in This Guide

Almost every Maple input and output you see in this guide appears exactly as we executed it in a Maple session. Slight changes were made only to enhance page layout or to guarantee better photo-offset production quality for graphics.

Also, each topic within the **Useful Tips** sections are denoted with *light bulbs* ♡ more light bulbs indicate tips that we think are more important than others. Our scale is 1 to 4 light bulbs, but your wattage may vary.

Final Comments

We would like to thank all of our colleagues and students at Boston College who have contributed to the completion of this guide. Special thanks are extended to Jenny Baglivo, Nancy Gaff, and Sarah Quebec for their contributions to the statistics pages in the first edition of the guide. Bill Keane contributed several suggestions on the penultimate printing of the guide. Bill Zahner was extremely helpful for his reading, rereading, and critiquing of (many versions of) the guide as it developed. He also offered terrific suggestions on presentation and arrangement.

The staff at John Wiley, Inc., has been very supportive of our efforts, as well as patient with our schedule which always seemed to be slightly missing those deadlines we expected to meet. Our sincerest thanks go to our editor David Dietz and his staff Pamela Lashbrook and Shannon Corliss for their enthusiastic response to this project.

Finally, despite our best efforts, it is likely that somewhere in these many pages, an error of either omission or commission awaits you. We sincerely apologize if this is the case and accept full responsibility for any inaccuracies.

We will provide Maple worksheets with the examples from the text at our websites. We will provide a listing of further comments, examples, and any inaccuracies that may be found in this Guide. You can locate us and our websites by visiting our departments on the web at either:

`www.bc.edu/math` *or* `math.slu.edu/~may` or `www.sc.edu/~meade`

We wish you only the best computing experiences with Maple.

Doug, Mike, C-K, and Jerry

Table of Contents

Part I. Basic Maple Commands

Part II. Drawing Pictures in Maple

Part III. Maple for One Variable Calculus

Part IV. Maple for Multivariable Calculus

Part V. Maple for Linear Algebra and Vector Calculus

Part VI. Using Maple in Statistics

Part VII. Advanced Features of Maple

Appendices and Index

Running Maple

Computer Systems

What Computer System Are You Using?

Maple software runs on almost every major computer system including mainframes and desktop systems. Maple can also be set up to run across a network and even between systems. Almost all implementations of Maple are graphically-based systems where you can both type on a keyboard and use a mouse to navigate a window (e.g., desktop systems running MS Windows, the MacOS or a Linux-based system.) The Maple commands we discuss in this guide will work on all of these implementations.

> **Note:** Text-based implementations of Maple are also available. These are often found on mainframe systems, and are usually of interest only to people who are doing remote logins, or to advanced users who want the performance advantage that comes from using a text-based system.

Starting the Software

You should follow the instructions that came with the Maple software to install it on your computer system. Once you've completed the installation, you're ready to explore Maple.

Starting Maple obviously depends on the system you are using.

- On most PCs and Macintosh computers, you will typically find the icon of the Maple application in a window. Click (or double-click) on the icon.

- On Linux, or other command-line operating systems, you will usually enter the command **xmaple** or **maple** (or a local equivalent, depending on how the software has been installed and how your system is configured).

> **Note:** If you run Maple over a network, you may need to check with your system manager for the starting procedure.

If this is the first time you start Maple, you will be presented with the choice to open a document or a worksheet. In the document mode, you can mix Maple's computation with your text to create an interactive document. In the worksheet mode, Maple's input has to be entered line by line in groups. (Please see the graphics of these two modes on the next page.)

Document is the default mode, but we find that beginners are more comfortable with the worksheet mode. Nevertheless, the Maple input we discussed in this book will work on both modes. If you want to switch from one mode to the other, check the Q & A section at the end of this chapter.

When Maple is launched, a new window will be opened with the cursor flashing, awaiting your input.

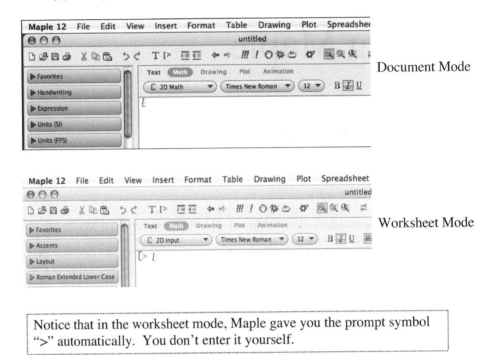

Document Mode

Worksheet Mode

Notice that in the worksheet mode, Maple gave you the prompt symbol ">" automatically. You don't enter it yourself.

Input and Output

Maple is interactive software. For almost every entry you make, Maple will provide a direct response. Once you launch Maple and see the flashing cursor, you can type the three characters $\boxed{1+1}$ and then press the evaluation key (either the $\boxed{enter}$ or $\boxed{return}$ key, depending on your computer system). Maple will give you the response "2" on a new line, and your window should look something like this:

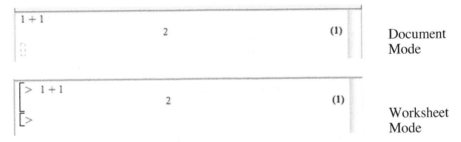

Document Mode

Worksheet Mode

Now you enter the characters $\boxed{39 - 11}$ and then press the evaluation key again ($\boxed{enter}$ or $\boxed{return}$) and you'll see the result of 28:

Document Mode

Worksheet
Mode

Throughout the rest of this manual, we will not show you windows. Instead, we'll show all of our input, with the response in plain text. Thus, our sequence above would be listed as:

Keystrokes: **1+1** $\boxed{enter}$

1 + 1

$$2$$

Keystrokes: **39–11** $\boxed{enter}$

39 – 11

$$28$$

Quit

When you've had enough and want to exit Maple, you can simply choosing the **Quit Maple** item in the **Maple** menu on the top of the screen. On a text based system, simply execute

```
quit
```

2-D Math Input

By default, Maple opens in 2-D Math input mode in a document. This allows you to enter the input in the same way as you would write it mathematically. As a result, you can use Maple with a minimum of syntax and mix mathematical notation with text for explanation. For example, to compute $\frac{2}{3}+1$, as you begin to type

2/

the input on the screen will automatically turn to the fraction form as shown below with a flashing cursor at the denominator waiting for your input.

$$\frac{2}{}$$

You can now enter **3** in the denominator. Press $\boxed{\rightarrow}$ (the right arrow key) on your keyboard and the cursor returns to the central position. Now type **+1**. You will see

$$\frac{2}{3}+1$$

Press the evaluation key $\boxed{enter}$ and Maple responds with

$$\frac{5}{3}$$

As another example, to enter $x^2 + 3x$ in 2-D Math mode, you will type **x^2** then press $\boxed{\rightarrow}$, and finish the typing with **+3*x** to see the input:

$$x^2 + 3 \cdot x$$

(When you type ^, the cursor will move to the superscript position. Press $\boxed{\rightarrow}$ to make the cursor move out of the superscript position.)

A Quick Tour

For the rest of this chapter, we'll show you some of Maple's capabilities. We present these examples only to whet your appetite. You can follow along at your computer by typing (exactly!) what we show below. In later chapters, we'll give you a more complete explanation of how to use these commands and how to use palettes to graphically produce the commands.

Note: When you input the following commands in Maple, make sure that:

- You use upper- and lower-case characters exactly as we do. Maple is very "case sensitive." If you use the wrong capitalization, Maple won't understand what you mean.

- Use exactly the type of brackets we show. There are three types of brackets: **[** *square brackets* **]**, **(** *parentheses* **)** and **{** *curly braces* **}**. Each has its own meaning in Maple. If you use the wrong one, Maple will be confused.

- You can continue input from one line to another in Maple by pressing either $\boxed{shift\text{-}return}$ or $\boxed{return}$ (depending on your system). For our examples, we recommend that you break lines exactly as we show in the text.

- You must always press the evaluation key $\boxed{enter}$ to see the output.

- Your output might appear slightly different from ours in some of the following examples. We explain why in Q & A at the end of this chapter.

Calculator   Maple does all the work of a hand-held electronic calculator. You can enter numerical expressions and Maple will do the arithmetic:

Keystrokes: **235.567*441.235/623.45** $\boxed{enter}$

$$\frac{235.567 \cdot 441.235}{623.45}$$

$$166.7181092$$

Keystrokes: **sin(0.3)** $\boxed{enter}$

$$\sin(0.3)$$

$$0.2955202067$$

But Maple can go much further. Try this factorial computation!

Keystrokes: **289!** enter

289!

2079866075306145164348895732262527092227125189083652864966524223171\
4057602959306387764301098263545191326756604339313630559109638714531\
7723797549314447666527391923032017635887236183475937403854287258461\
1225722710498189168763234932439760233029166663945402474493070106651\
7313319035568964279626035832919320283513188887861286895384890867131\
0054499895916955854460148058813107716107435786968970196238825729571\
3115726033710483762553382305725385845880790786699431748508548589951\
5805949142445628564109186070285204106844632108658746982400000000001\
0000000000000000000000000000000000000000000000000000000000000000000

Your calculator cannot do this!

Solving Equations

Maple can solve complicated equations and even systems of equations in many variables. For example, the equations $2x + 5y = 37$ and $x - 3y = 21$ have a simultaneous solution:

Keystrokes: **solve({2*x+5*y=37,x-3*y=21},{x,y})** enter

$$solve(\{2 \cdot x + 5 \cdot y = 37, x - 3 \cdot y = 21\}, \{x, y\})$$

$$\left\{ x = \frac{216}{11}, \ y = \frac{-5}{11} \right\}$$

Maple can also find solutions to equations numerically. For example, the equation $x = \cos(x)$ has a solution very close to $x = 0.75$. We can find it with:

Keystrokes: **fsolve(x=cos(x),x)** enter

$$fsolve(x = \cos(x), x)$$

$$0.7390851332$$

You will learn more about solving equations in Chapter 6.

Algebra

Maple is very good at algebra. It can work with polynomials:

Keystrokes: **expand((x-2)^2 → *(x+5)^3 →)** enter

$$expand((x - 2)^2 \cdot (x + 5)^3)$$

$$x^5 + 11x^4 + 19x^3 - 115x^2 - 200x + 500$$

> **Note:** Use → after the **2** to move out of the superscript. The → will also move you out of the denominator of a fraction or the radical of a square root.

Keystrokes: `factor(x^5 → +11*x^4 → +19*x^3 → -115*x^2 → -`
`200*x+500) enter`

$$factor(x^5 + 11 \cdot x^4 + 19 \cdot x^3 - 115 \cdot x^2 - 200 \cdot x + 500)$$

$$(x-2)^2(x+5)^3$$

Are you impressed? Maple also knows standard trigonometric identities such as $\sin^2(x) + \cos^2(x) = 1$:

Keystrokes: `simplify(sin(x)^2 → +cos(x)^2 → ) enter`

$$simplify\left(\sin(x)^2 + \cos(x)^2\right)$$

$$1$$

> Note that $\sin(x)^2$ means $\left(\sin(x)\right)^2$ and is different from $\sin(x^2)$.

That one was easy, but you probably forgot that $\sec^2(x) - \tan^2(x) = 1$:

$$simplify\left(\sec(x)^2 - \tan(x)^2\right)$$

$$1$$

You will learn more about doing algebra in Maple in Chapter 5.

Calculus

Maple even knows a lot about calculus! We can find the derivative of the function $f(x) = x / (1 + x^2)$.

Keystrokes: `diff(x/(1+x^2 → ) → ,x) enter`

$$diff\left(\frac{x}{(1+x^2)}, x\right)$$

$$\frac{1}{1+x^2} - 2\frac{x^2}{(1+x^2)^2}$$

A complicated integral such as $\int \frac{1}{1+x^3}\, dx$ is handled rather easily.

Keystrokes: `int(1/(1+x^3 → ) → ,x) enter`

$$int\left(\frac{1}{1+x^3}, x\right)$$

$$-\frac{1}{6}\ln(x^2 - x + 1) + \frac{1}{3}\sqrt{3}\arctan\left(\frac{1}{3}(2x-1)\sqrt{3}\right) + \frac{1}{3}\ln(1+x)$$

In Chapter 2, we will show you how to use palettes for differentiation and integration. Chapters 12 through 15 demonstrate many of the Calculus capabilities of Maple. The Appendix A also contains information on Maple commands useful for learning Calculus.

Graphing in the Plane

Maple does everything a standard graphing calculator does and does it better. For example, to see the graph of the function $f(x) = x/(1+x^2)$ over the interval $-4 \le x \le 4$, you can use:

Keystroke: `plot(x/1+x^2` $\boxed{\rightarrow}$ $\boxed{\rightarrow}$ `,x=-4..4)` $\boxed{enter}$

$$plot\left(\frac{x}{1+x^2}, x = -4..4\right)$$

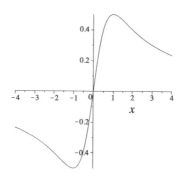

We can see a "daisy" with:

Keystroke: `plot([cos(21*t)*cos(t),cos(21*t)*sin(t),t=0..2*Pi])` $\boxed{enter}$

$$plot([\cos(21 \cdot t) \cdot \cos(t),\ \cos(21 \cdot t) \cdot \sin(t),\ t = 0..2 \cdot Pi])$$

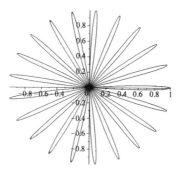

Chapters 9, 10, and 11 contain the details of the two-dimensional plotting capabilities of Maple.

Plotting in Space

Maple does a wonderful job with three-dimensional graphics. Let us show you two examples.

Keystrokes: `plot3d(sin(x)*cos(y),x=0..2*Pi, y=0..2*Pi)` $\boxed{enter}$

$$plot3d(\sin(x) \cdot \cos(y),\ x = 0..2 \cdot Pi,\ y = 0..2 \cdot Pi)$$

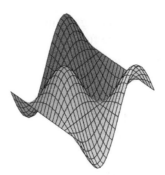

Keystrokes: **plot3d([t,r*cos(t),r*sin(t)],r=0..1,t=0..6*Pi,** $\boxed{shift\text{-}return}$

grid=[8,60]) $\boxed{enter}$

plot3d([t, r · cos(t), r · sin(t)], r = 0..1, t = 0..6 · Pi, grid = [8,60])

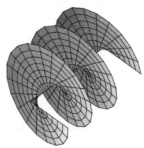

Chapters 16 and 20 contain most of the information you need to create three-dimensional pictures.

**Changing the
3-D View**

If you click and drag the mouse button with the cursor/arrow at any point inside a 3-D picture, the picture will rotate according to how you move the mouse. This allows you to see the 3-D picture from any viewpoint.

Chapter 20 contains more information on how you can utilize Maple's interface to more easily control your view of 3-D graphics.

**Programming
and
Simulation**

Maple has its own programming language. You can use it to write code just as you would in Java, C or BASIC. Here's a simple routine to simulate the flipping of a coin several times and return the number of heads observed:

Keystrokes: **coinflips := proc(howmany)** $\boxed{shift\text{-}return}$

    local heads; $\boxed{shift\text{-}return}$

    heads := 0; $\boxed{shift\text{-}return}$

    from 1 to howmany do $\boxed{shift\text{-}return}$

    if RandomTools[Generate](choose([head,tail])) $\boxed{shift\text{-}return}$

      =head then $\boxed{shift\text{-}return}$

        heads := heads + 1; $\boxed{shift\text{-}return}$

```
end if; shift - return
end do; shift - return
printf("%a heads seen in %a flips", shift - return
        heads, howmany ); shift - return
end proc; enter
```

You can use this routine:

Keystrokes: **coinflips(100);** enter
            55 heads seen in 100 flips.

Keystrokes: **coinflips(1000);** enter
            491 heads seen in 1000 flips.

Chapter 27 introduces the basic features of the Maple programming language.

More Examples

The "More Examples" sections of this guide present examples involving more mathematics. Students of mathematics, science, and engineering may find these of interest.

Here's one such example. Not too many people know about the Bessel functions. But if you're learning physics, you might want to know what Maple has available for you in Bessel functions.

Special Functions

The Bessel function of order 0, $J_0(x)$, is a solution to the differential equation:

$$x^2 y'' + xy' + x^2 y = 0 .$$

You can see its graph with:

Keystrokes: **plot(BesselJ(0,x),x=0..20)** enter

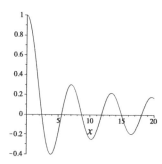

The smallest, positive zero of $J_0(x)$ is at approximately $x = 2.40483$:

Keystrokes: **fsolve(BesselJ(0,x)=0,x=1)** enter
            2.404825558

Useful Tips

 If you want to try the examples in this book, you don't have to enter the Maple input one by one by hand. We've made available copies of the Maple documents and worksheets used to create the inputs you see in this Guide. After downloading them, you can copy and paste into your own Maple document and begin experimenting. Please visit the authors' websites mentioned in the Preface.

Troubleshooting Q & A

Most chapters of this guide end with a Troubleshooting Q & A section, where we answer some common questions we think you will have. But there's not too much you can ask yet, except:

Question... Maple is giving me unexpected evaluations when I use trigonometric or other named functions in 2-D Math. What am I doing wrong?

Answer... A common mistake when entering 2-D Math input is to carelessly leave a space between the function name and the left parenthesis. In 2-D Math input a space is implied multiplication, but it may be hard to see. For example:

Keystrokes: **sin(1.2)** $\boxed{enter}$

sin(1.2)

0.9320390860

But if you type a space, then the answer will be very wrong.

Keystrokes: **sin** $\boxed{space}$ **(1.2)** $\boxed{enter}$

sin (1.2)

1.2 sin

Question... Suppose I make a mistake in the input, or I do not finish the entry, but I hit $\boxed{enter}$ anyway. How will Maple respond?

Answer... Most likely you will get an error message from Maple. Don't worry. You can simply move the cursor back to your input, correct the mistake or continue the typing to finish the input.

Question... The output I get when testing some of the examples in this text does not look exactly like what's shown here. What's going on?

Answer... All the outputs we shown in the text are from the document mode of Maple 12. If you use a different version of Maple, or use the worksheet mode instead, you may get an output in a different format or arrangement. However despite this inconsistency, the actual computation and the final result should not be affected.

Question...   If I set up the document mode as the default, how can I start Maple with worksheet mode instead, or vice versa?

Answer...   No matter whether you are in a document mode or worksheet mode, you can always open a new document or worksheet by choosing **File → New → Worksheet Mode** or **Document Mode** from the menu bar.

You can also reset the default style of user interface that Maple starts by default. To do this, click on the **Tools** menu and select the **Options ...** entry from the list. A popup window will appear. Now, click on the **Interface** tab and look for the drop-down list labeled by "**Default format for new worksheets:**". Select **worksheet** or **document**, then click **Apply Globally**.

Question...   What do I do now?

Answer...   Turn the page and start learning about Maple!

CHAPTER 2
Maple Input Modes

2-D Input Mode

2-D Math Input   By default, when you open Maple document or worksheet, it is ready for 2-D input. As we have seen in the examples in the first chapter, 2-D input allows us to enter and automatically format a mathematical expression in the same way as we would write it by hand. Here is a short summary of some of the special keys used for 2-D input:

Key(s)	What it does	Example
^	Exponent	*Keystrokes*: **3^2** The input appears as: 3^2
/	fraction	*Keystrokes*: **3/1+x** The input appears as: $\dfrac{3}{1+x}$
*	Multiplication	*Keystrokes*: **x*y** The input appears as: $x \cdot y$
$\boxed{\rightarrow}$ (Right arrow key)	Navigating expressions	*Keystrokes*: **3^2** $\boxed{\rightarrow}$ **x** The input appears as: $3^2 x$

> **Note**: Although one can also use the $\boxed{\text{space}}$ (space bar) in place of * for multiplication, we do not recommend it because Maple doesn't automatically insert a multiplication symbol. It is difficult for a novice to pick up the difference between the inputs $x\,y$ (which means x times y) and xy (which means the name xy).

■ **Example.** To enter $\dfrac{x^2}{1+x} + 5$ with 2-D Math input, we type:

Keystrokes: **x^2** $\boxed{\rightarrow}$ **/1+x** $\boxed{\rightarrow}$ **+5**

Palettes Input   In addition to directly entering an expression using only keystrokes, you can use pre-configured palettes to format and organize the terms of your input. By default, more than 20 palette docks are displayed in the left pane of the Maple document.

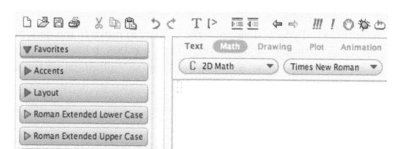

(If you don't see the palettes, you can load them from the main menu by selecting **View → Palettes → Show All Palettes**.)

The most frequently used palettes are the **Expression**, **Matrix**, and **Common Symbols**. When you click on a palette dock, all of the templates, symbols, and other items inside that particular palette will be displayed. You can insert a template into your session by pointing the cursor at the icon and clicking the left mouse button.

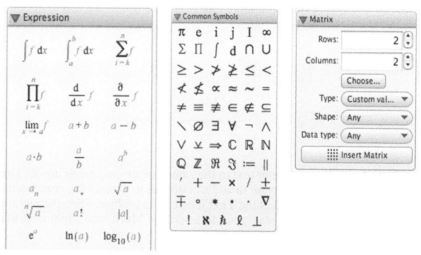

By using templates from the **Expression** Palette, we can easily ask Maple to integrate, differentiate, find limits and summations, and so on. We will quickly illustrate how they work with the following example.

■ **Example**. To evaluate $\int \dfrac{1}{2 + \pi x^2}\, dx$ using palettes:

• Open the **Expression** palette by clicking on its button. Click on the indefinite integral template ($\boxed{\int f\, dx}$) in the **Expression** palette. A new expression is created in your document. Notice that f is highlighted in the expression.

$$\int f\, dx$$

• Click on the fraction template ($\boxed{\dfrac{a}{b}}$) in the **Expression** palette. Now you see:

$$\int \frac{1}{b} \, dx$$

- Type **1**, then press the *tab key* to move to the denominator. (Every time you press the *tab key*, you move from one box in the expression to another box.) Then type **2 +** , and then click on the multiplication template ($\boxed{a \cdot b}$). Now you see the expression:

$$\int \frac{1}{2 + \boxed{a \cdot b}} \, dx$$

- Open the **Common Symbols** palette. Click on the $\boxed{\pi}$ symbol in that palette. Then press the *tab key*.

$$\int \frac{1}{2 + \pi \cdot \boxed{b}} \, dx$$

- Click on the power template ($\boxed{a^b}$) in the **Expression** palette. Type **x**, press the *tab* key, then type **2**, and press the *tab* key again. Now you see the completed expression more like what you would see in a textbook. Hit $\boxed{enter}$ to see Maple's output.

$$\int \frac{1}{2 + \pi \cdot x^2} \, dx$$

$$\frac{1}{2} \frac{\sqrt{2} \arctan\left(\frac{1}{2}\sqrt{\pi} \; x \; \sqrt{2}\right)}{\sqrt{\pi}}$$

Adding Text to Math Input

You can enter text in the same line as your math input. This is particularly helpful when you want to add a comment to explain the computation. In order to do so, you have to select the **Text** icon on the toolbar. (See the picture below.) After you finish typing the text, you can select the **Math** icon (right next to the **Text** icon) on the tool bar to continue the Math input.

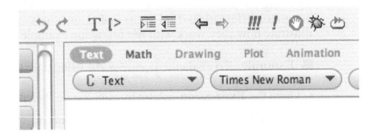

- **Example**. We want to comment that the area is found by computing the integral $\int_0^2 x \, dx$. The procedure is:

  - Select the **Text** icon on the toolbar. Type **The area can be computed with**.

- Select the **Math** icon on the toolbar. From the **Expression** palette, select the definite integral template ($\boxed{\int_a^b f\,\mathrm{dx}}$).

- Notice that the placeholder for *a* is highlighted. Type **0**, then press the *tab* key, type **2**, press the *tab* key again, and then type **x**.

- Press the evaluation key (**Return/Enter**). We have:

The area can be computed with $\displaystyle\int_0^2 x\ dx$

2

1-D Input Mode

1-D Math Mode   You can also enter input in 1-D Math mode, which Maple refers to as Maple Notation or Maple Input. This is the default mode for older versions of Maple, where no automatic formatting of input was available. Many Maple users still prefer to use this input mode; this is particularly true amongst users who write Maple programs.

If you want to switch from the default 2-D mode into 1-D input mode, you have to first select **Maple Input** from the **Insert** menu at the menu bar. (Or you can use **command-M** as a keyboard equivalent.) Now, the slanted flashing cursor in the document changes to a vertical cursor. You can type, for example, **2/3;** and then hit the evaluation key (**Return** or **Enter**). You will get

    2/3;

$$\frac{2}{3}$$

> Notice that with 1-D Math input,
>
> - input appears in red and in a fixed-width font;
>
> - Maple does not automatically format your input with a numerator and denominator as you type; and
>
> - all 1-D input must end with a semicolon.

Comparison of 1-D and 2-D Input Modes

You can intermix the use of 1-D and 2-D math modes throughout a Maple document. Each mode has its advantages. Nevertheless, some users will be strong advocates of using one mode over the other. So it helps to compare.

Comparing Keystrokes

■ **Example**. We want to enter the expression $\dfrac{1}{1+x^2}+5$ into Maple.

• Using the 2-D Math input

Keystrokes: **1/1+x^2** $\boxed{\rightarrow}$ $\boxed{\rightarrow}$ **+5**

(The first $\boxed{\rightarrow}$ brings the cursor out of the exponent position, and the second $\boxed{\rightarrow}$ brings us out of the denominator.)

• Using the 1-D Math input

Keystrokes: **1/(1+x^2) + 5;**

Notice that in the 1-D Math input, we have to use parentheses to group the terms in the denominator together. If you enter instead

Keystrokes: **1/1+x^2 + 5;**

Maple will misinterpret your input as $\dfrac{1}{1}+x^2+5$, and return $6+x^2$.

Converting Between Input Modes

You can convert between 1-D and 2-D Math by highlighting the input to be converted and choosing **Convert To** from the **Format** menu, then selecting between the **1-D Math input** option or the **2-D Math input** option.

For example, you can convert 2-D Math inputs,

$$\frac{2}{3}+1$$

and

$$x^2+3\cdot x$$

into 1-D Math inputs

2/3+1;

x^2+3*x;

respectively.

Which Mode To Choose?

Now, let us give you a simple, direct comparison between the 2-D Math input and 1-D input.

	2-D Input	1-D Input
Advantages	• Easier for beginners because it matches how mathematics is typically presented in textbooks. • Input errors are easier to identify because you see the expression as you type.	• Looks like calculator syntax. • Easier for users to replicate because every keystroke is recorded.

Disadvantages	• Subtle differences in the output can lead to confusion. For example, the 2-D inputs "*a b*" and "*ab*" have very different interpretations. • The expressions "sin (Pi/2)" and "sin(Pi/2)" also illustrate another error that is difficult to detect. • Some mistakes cannot be undone without completely restarting the expression.	• Matching parentheses is essential. Maple's built-in parenthesis matching is helpful. • All groupings must be done with parentheses; [*square brackets*], {*curly braces*}, and (*parentheses*) have different uses in Maple. (See Chapter 6.) • When entering a complicated math expression, users may sometimes fail to group some of the terms correctly or leave off a matching parenthesis and hence get an error message.
Differences in keystrokes	• $\boxed{\rightarrow}$ is used to move the cursor within the expression. • $\boxed{\text{space}}$ means (implied) multiplication, but we discourage its use.	• Use $\boxed{\rightarrow}$ to move cursor to the right. • $\boxed{\text{space}}$ has no special meaning. • Must end an input with : or ; • Parentheses are required to correctly group terms.

Which Input Method Will This Book Represent?

Each input method has its own advantage and disadvantage. In the first few chapters of this book, we will show you both the 2-D and 1-D inputs, but later on we use only the traditional 1-D Maple input. Throughout this manual, when you see, for example,

Keystrokes: **3^2** $\boxed{\rightarrow}$ **/5**

$$\frac{3^2}{5}$$

The first line shows the keystrokes used to enter the expression. This is followed by the actual 2-D Math input display. On the other hand, if you see

3^2/5;

that is the input for 1-D Math.

Readers should feel free to use either method, but be ready to convert to 1-D if Maple gives an unexpected result. In the other direction, converting to 2-D is often useful to check whether the Maple command is asking for what the user hopes to compute.

Useful Tips

    In 2-D input, do not use spaces for implicit multiplication. Instead, use * explicitly.

☿ ☿   When working with palettes, the *tab* key moves through the input areas of the current template.

☿   Make a habit of using **:** or **;** to end a Maple input in both 2-D and 1-D input. This will avoid unnecessary warnings and error messages during input.

Troubleshooting Q & A

Question...   How come $\boxed{\rightarrow}$ (the right-arrow-key) does not work in moving the cursor from one field to the other in 2-D input?

Answer...   In regular 2-D Math input, you use $\boxed{\rightarrow}$ to move the cursor about the various terms in the expression. To move the cursor about the terms of a palette, you use the *tab key*. In both modes, you can use the mouse to move the cursor directly to the location within the expression where you want to insert (or change) the input.

Question...   When I convert an expression to 1-D, Maple adds a semicolon to my expression. Why?

Answer...   1-D Math was designed for text based systems and set up to have no ambiguities. A semicolon indicates both the end of an expression and the instruction that Maple should evaluate the expression and report back on its value.

CHAPTER 3
Calculator Features

Simple Arithmetic

Basic Arithmetic Operations

In this chapter, we will show you how Maple can work as a calculator. We begin with the basic arithmetic operations used in Maple:

Command	What It Does
+, −	Add, Subtract
*, /	Multiply, Divide
^	Raise to a power (exponentiation)

Here is an example to evaluate $\dfrac{23}{5} - \dfrac{3}{5} + 5(2^3)$ with either 2-D or 1-D input:

Keystrokes: **23/5** $\boxed{\rightarrow}$ **−3/5** $\boxed{\rightarrow}$ **+5*2^3** $\boxed{\rightarrow}$

44

> **Note:** $\boxed{\rightarrow}$ is the right-arrow-key on the keyboard, we can use it to get out of the denominator and the exponent as discussed in Chapter 2.

23/5−3/5+5*2^3;

44

We can also use parentheses to group terms together. For example, the expression $(3+4)\left(\dfrac{4-8}{5}\right)$ can be entered in 2-D or 1-D Math mode with:

Keystrokes: **(3+4)*((4−8)/5** $\boxed{\rightarrow}$ **)**

(3+4)*((4−8)/5);

$$-\frac{28}{5}$$

> **Note:** Use **(** *parentheses* **)** to group terms in expressions. Do not use **[** *square brackets* **]** or **{** *curly braces* **}** -- they mean something different. (See Chapter 6.)

Precedence   In 1-D Math input, Maple follows the laws of precedence of multiplication over addition and so on, just as you do by hand. For example,

> **9/6*22+5;**
>
> > 38

actually computes $(\frac{9}{6} \times 22) + 5$. A common mistake is to think of the input as either $\frac{9}{6 \times 22} + 5$ or $\frac{9}{6} \times (22 + 5)$. In 2-D Math input, Maple follows the visual grouping. This means you see exactly the expression that Maple actually computes.

Comments   You can add a comment to any expression by starting the comment with the pound sign **#**. For example in 2-D Math mode,

Keystrokes: **27*3 # This multiplies 27 and 3.**

> **27·3**   *# This multiplies 27 and 3.*
>
> > 81

Or in 1-D mode, we enter this as the following (and don't forget to add the semicolon).

> **27*3; # This multiplies 27 and 3.**
>
> > 81

Maple will ignore the phrase **This multiplies 27 and 3**. It is for your own reference.

We'll use comment lines throughout this guide to write short reminders about what's being emphasized in certain examples.

Previous Results and % Syntax   As a convenience, Maple lets you use the percentage sign, **%**, to stand for "the last result obtained." In this way, you can avoid retyping output when you want to work with it in your next input.

For example,

Keystrokes: **3200*12**

> **3200·12**
>
> > 38400

Keystrokes: **% - 6500**   **# Here % refers to 38400**

> **% – 6500**   *# Here % refers to 38400.*
>
> > 31900

In 1-D Math input mode, we can enter:

> **(1000 + %)^2;**   **# Here % refers to 31900.**
>
> > 1082410000

You can use two percentage signs **%%** for the "second-last result obtained," and three percentage signs **%%%** for the third-last. However, this convention does not work with more than 3 percentage signs.

You can also refer to specific output by its equation label. (The equation label is the number in parentheses at the right-hand edge of each output.) To do that you can type **CTRL+L** (or **Command+ L** for Mac). An **Insert Label** dialog is displayed (see the picture below). Enter the label number and click OK.

Output Styles

Numeric Output

Maple gives you an exact (symbolic) value for almost every numeric expression:

    (3+9)*(4-8)/1247*67;   # for 1-D mode, or

$$\frac{(3+9)\cdot(4-8)}{1247}\cdot 67 \qquad \textit{# for 2-D mode}$$

$$-\frac{3216}{1247}$$

You can force Maple to give you an answer that looks like the decimal answer you'd get on a calculator by using **evalf** (for floating point evaluation) with parentheses around an expression:

    evalf((3+9)*(4-8)/1247*67);   #for 1-D mode, or

$$evalf\left(\frac{(3+9)\cdot(4-8)}{1247}\cdot 67\right) \qquad \textit{#for 2-D mode}$$

$$-2.578989575$$

By default Maple shows floating-point answers with 10 digits (including those to the left of the decimal point). You can see more digits in the answer – say 40 – with:

    evalf((3+9)*(4-8)/1247*67, 40);   # for 1-D mode, or

$$evalf\left(\frac{(3+9)\cdot(4-8)}{1247}\cdot 67, 40\right) \qquad \textit{# for 2-D mode}$$

$$-2.578989574979951884522854851643945469126$$

Results like these are called **approximate numeric values** in Maple.

Scientific Notation

Maple uses a modified standard scientific notation to display results when the numbers either get very large or very small:

    evalf(1234567890);   # for both 1-D and 2-D modes

$$1.234567890 \ 10^9$$

0.0000003492836;   # for both 1-D and 2-D modes

$$3.492836 \ 10^{-7}$$

The spaces in the output above represent (implied) multiplication.

Built-in Constants and Functions

Built-in Constants

The mathematical constants used most often are already built into Maple. You can either use the **Common Symbols** Palette (see Chapter 2), or type them in directly. Be careful using upper-case and lower-case characters when you use these constants.

Constant	Value	Explanation	Maple
π	3.1415926...	Ratio of a circle's circumference to its diameter	`Pi`
e	2.71828...	Natural exponential	`exp(1)`
i	$i = \sqrt{-1}$	Imaginary unit, a solution to x^2+1=0	`I`
∞	∞	(Positive) infinity	`infinity`

For example, to see the value of π to 45 significant digits, we use:

evalf(Pi, 45);   # for both 1-D mode and 2-D mode

3.14159265358979323846264338327950288419716940

To compute the numerical value of $\pi^4 - 5e$, we type:

evalf(Pi^4 −5*exp(1));   # for 1-D mode, or

$evalf\left(\pi^4 - 5 \cdot \exp(1)\right)$ # **for 2-D mode**

83.81768194

Built-in Functions

Maple has many built-in functions. Here are the functions that you will probably use the most. (Some of these can be found in the **Expression** Palette.)

Function(s)	Sample(s)	Maple Name(s)
Natural logarithm	$\ln(x)$	`ln(x)`
Logarithm to base a	$\log_a x$	`log[a](x)`
Exponential	e^x	`exp(x)`
Absolute value	$|x|$	`abs(x)`
Square root	$\sqrt{x}$	`sqrt(x)`

Trigonometric	$\sin(x)$, $\cos(x)$, ...	`sin(x),cos(x),tan(x),` `cot(x),sec(x),csc(x)`
Inverse trigonometric	$\sin^{-1}(x)$, $\cos^{-1}(x)$, ...	`arcsin(x),arccos(x),` `arctan(x),arccot(x)`, etc.
Hyperbolic	$\sinh(x)$, $\cosh(x)$, ...	`sinh(x),cosh(x),` `tanh(x),coth(x),` `sech(x),csch(x)`
Inverse hyperbolic	$\sinh^{-1}(x)$, $\cosh^{-1}(x)$, ...	`arcsinh(x),arccosh(x),` `arctanh(x),arccoth(x),` etc.

For example,

`sin(Pi);`

$$0$$

`evalf(sin(180));`   # 180 *radians, NOT* 180 *degrees.*

$$-.8011526357$$

`arctan(1);`

$$\frac{1}{4}\pi$$

`exp(ln(exp(1)));`

$$\mathbf{e}$$

Notes:

(1) Maple uses radian measure for all angles. To convert from degrees to radians, multiply by `Pi/180`.

(2) Maple refers to e^1 as **e** in the output. Unfortunately, you cannot use **e** (bold face letter e) when you enter input. You have to write `exp(1)` for input.

(3) A common mistake for the beginner is to enter `e^x` for the exponential function e^x, it has to be `exp(x)` instead.

Error Messages

If you make a mistake in your input, Maple *displays an error message.* In 2-D Math input, Maple will also put a red box around the part of the output where the mistake was detected. In 1-D Math input, Maple places the cursor on the position where the error was detected. Note that the actual error can be elsewhere in the expression.

Common Mistakes

The most common mistakes for beginners usually involve mismatching (or omitting) parentheses and forgetting to write the multiplication symbol, as you see in the following examples:

```
3·(4 − 5))+ 6 # too many right parentheses
Error, unable to match delimiters

sin 3  # it should be sin(3)
Error, missing operation

5 + 2^3  # it should be 5 + 2^x·3
Error, missing operation
```

$3 \cdot (4 − 5)) + 6$ # too many right parentheses

$\sin 3$ # it should be sin(3)

$5 + 2^x\,3$ # it should be $5 + 2^x \cdot 3$

More Examples

Approximate Numbers and Exactness

Working with **approximate numeric values** is just like working with values on a hand-held calculator. These numbers sometimes lose precision as values get rounded off in arithmetic.

■ **Example**. Maple makes a big distinction between the exact number π and a numerical approximation for it. For example, **sin(517*Pi)** is an exact quantity with an exact answer:

sin(517*Pi);   #for 1-D mode, or

$\sin(517 \cdot \mathbf{Pi})$ *#for 2-D mode*

$$0$$

If you use a numerical approximation for 517π, you don't get an exact zero:

sin(evalf(517*Pi));   #for 1-D mode, or

$\sin(evalf(517 \cdot \mathbf{Pi}))$ *#for 2-D mode*

$$-9.407689571 \ 10^{-8}$$

This is very close to zero (it is, after all, -0.00000009407689571) and it's probably acceptable for the work you will be doing. But it's not exact.

Useful Tips

Don't forget to type the asterisk (*****) when multiplying terms. This is the most common mistake made by beginners. For example, you might incorrectly enter **cos(0)23** instead of **cos(0)*23** and get an error message. However, *in 2-D Math Input mode*, if a number comes before a function, this will be interpreted as multiplication. But, if the same input is entered using 1-D Math Input, Maple considers this to be an error.

For example, in 2-D Math Input mode, **23cos(0)** is interpret as 23*cos(0), but in 1-D mode **23cos(0);** produce an error message that states "missing operator or \`;\`." Because of this confusion, it is better to use * explicitly for multiplication.

☽ ☽ ☽ ☽ Never try to multiply terms together in Maple using only **(** *parentheses* **)**. For example, in mathematical writing you can write (cos(Pi))(4) to denote $\cos(\pi) \times 4$. However, Maple has a different interpretation.

(cos(Pi))(4)

-1

☽ ☽ ☽ Use parentheses in expressions to clarify what you mean. This helps avoid mistakes. For example, you might think that **3^2*x** means 3^{2x}, but it doesn't! It is actually $(3^2)x$ because the square is done before the multiplication of **2** and **x**. To get 3^{2x}, you should write **3^(2*x)**.

☽ ☽ Many of the Maple built-in functions (e.g., **sin**, **cos**, **ln**) are understood to be defined for *complex* arguments. As a result, you may sometimes get an unexpected result that involves a complex number.

☽ ☽ Use the ditto operator (**%**) sparingly. We recommend using the **%** symbol only for short sequences of one or two calculations that you don't expect to repeat later. In general, it is better to use equation labels when you need to refer to a previous result,

☽ Avoid using **exp(1)** together with ^ to describe an exponential function. For example, the expression e^{2x} is better written as **exp(2*x)**, rather than **exp(1)^(2*x)**. (The latter expression has to be simplified before Maple recognizes it as being the same as **exp(2*x)**.)

Troubleshooting Q & A

Question... When I tried to evaluate a built-in function, Maple gave an error message, "Error, unexpected number." What should I check?

Answer... Make sure that you included parentheses when using Maple functions. For example, a common mistake for beginners is to type:

sin 2;

Error, unexpected number

The correct input should be **sin(2);**.

Question... When I entered **sin(2)**, Maple just gave me the same thing back again. It didn't evaluate it. Why?

```
sin(2);
```

$$\sin(2)$$

Answer... Maple always gives you an exact answer. When you write sin(2) in mathematics, you don't try to simplify it. Neither does Maple. However, if you want a decimal approximation for sin(2), use **evalf**.

```
evalf(sin(2));
```

$$0.9092974268$$

Question... I cannot get a numerical value of π from Maple using **evalf**. Why?

Answer... It is a common mistake to enter π as **pi** instead of **Pi**. Maple will consider **pi** as the Greek letter "π" rather than the geometrical constant "π". Therefore, **pi** does not have any numerical value.

```
pi, evalf(pi), Pi, evalf(Pi);
```

$$\pi,\ \pi,\ \pi,\ 3.141592654$$

Question... When I tried to evaluate a built-in function, Maple just returned the input unevaluated. I then used **evalf**, but still Maple did not give me a numerical value. What should I check?

Answer... Check your spelling. Most likely, you've misspelled the name of a built-in function or constant. For example, you may have used **cosine(Pi)** or **cos(pi)** instead of **cos(Pi)**.

Question... I got the error message "Error, missing operator or `;`" when I entered a complicated arithmetic expression. What should I check?

Answer... There are many possible errors that will generate this error message. You should:

- Check the spelling of all of the commands and variables.

- Check if all the (*parentheses*) are in the right places. In particular, make sure you correctly matched right and left parentheses.

- Check that you typed * for multiplication. For example:

```
sin(5)3 + 7;
```
Error, missing operator or `;`

Instead, use.

```
sin(5)*3 + 7;
```

CHAPTER 4
Variables and Functions

Variables

Immediate Assignment

With a calculator, you can store a value into memory and recall it later. With a more advanced calculator, you can store several, different values under names such as A, B, C, and so on, and then use one or more of them in later calculations.

You can do even better in Maple. You can assign a name to any Maple expression or value and then recall it whenever you want. You do this using a colon followed by an equal sign, **:=**, that is the symbol for **immediate assignment**. (Note that there is no space between the colon and equal sign.) For example:

```
a := 3.4;                    # We assign a to have the value 3.4.
```
$$3.4 \qquad \text{# Output in the document mode.}$$
$$a := 3.4 \qquad \text{# Output in the worksheet mode.}$$

> Please note that from now on, in order to avoid repetition, we will only show you the 1-D input, in cases where the same keystrokes can be applied to the 2-D input. Also, we will only show you the output in the document mode.

Once you've made an assignment, you can recall its value or use it in an expression:

```
a;
```
$$3.4$$

```
a+2;
```
$$5.4$$

```
a^2;                         #You will see a² in the case of 2-D mode.
```
$$11.56$$

How Expressions Are Evaluated

You may want to know how Maple keeps track of all the symbols and variables that you have defined.

Say we assign the name **myLunch** to the sum of **apple** and 3 times **banana**.

```
myLunch := apple + 3*banana;
```
$$apple + 3\,banana$$

Maple spits back the definition because **apple** and **banana** have no associated values. Now suppose we give the value 2 to **apple** and then reevaluate **myLunch**:

```
apple := 2;                  # Now apple has the value 2.
```

2

myLunch;

$$2 + 3\ banana$$

(When Maple reevaluates **myLunch,** it substitutes the value 2 for **apple.**)

If we now define the value of **banana** to be 3 and reevaluate **myLunch:**

banana := 3;   # Now **banana** has the value 3.

$$3$$

myLunch;

$$11$$

When Maple reevaluates **myLunch**, it replaces **apple** and **banana** by their respective values, and simplifies the resulting expression to obtain 11. Bon appétit!!

Redefining and Clearing Symbols

Once a name has been assigned a value or an expression, Maple retains the association until you redefine it, explicitly clear it, or end your session.

a := 3.4;   # We assign **a** to be 3.4.

$$3.4$$

a := 5;   # We reassign **a** to be 5.

$$5$$

a + 2;   # Maple uses the most recently assigned value of **a**.

$$7$$

The easiest way to ask Maple to forget about an assignment is to use the **unassign** command.

unassign('a');   # Note that we use single quotes around **a**.
a;

$$a$$

You can also **unassign** assignments for several names at the same time, using one single statement:

unassign('myLunch','apple','banana');

Rules for Names

Names you use can be made up of letters and numbers, subject to the following two rules:

- You can't use a name that begins with a number. For example, **2app** is not an acceptable name because in 1-D input, Maple will think that you forgot to type an operator between "**2**" and "**app**," and so will give you an error message. In 2-D input mode, Maple will think that you mean "**2**" times "**app**."

- You can't choose names that conflict with Maple's own names. For example, you can't name one of your own variables **sin**.

All of the following are examples of legitimate names that you could use:

a, m, p1, A, area, Perimeter, Batman, classsOf2012

> **Note:** Maple distinguishes upper case and lower case characters. For example, the names **Batman**, **batman**, and **batMan** are different.

Evaluation Command

You can substitute values into an expression without defining the variables explicitly. The substitution command, **eval**, is used in the form:

> **eval**(*expression, set of substitutions using =*)

For example, to substitute $x = 2$ and $y = 5$ into the expression $x^2 - 2xy$:

> **eval(x^2-2*x*y, {x=2, y=5 });**   #1-D input, or

> *eval*($x^2 - 2 \cdot x \cdot y, \{x = 2, y = 5\}$)   *# 2-D input*
>
> -16

One advantage of using the **eval** command is that the evaluation of a variable is temporary and is not remembered by Maple.

> **x;**   # The value of x is unchanged by the previous substitution.
>
> x

Finding the Larger Root of a Quadratic Equation

■ **Example**. We want to find the larger root of each of the following quadratic equations: $2x^2 + 5x - 6 = 0$ and $2x^2 + x - 3 = 0$.

The roots of a quadratic equation $ax^2 + bx + c = 0$, with $a \neq 0$, are found using the quadratic formula $(-b \pm \sqrt{b^2 - 4ac})/(2a)$. The larger root, when $a > 0$, is thus:

> **unassign('a','b','c','largerRoot');**
> **largerRoot := (-b+sqrt(b^2-4*a*c))/(2*a);**   # 1-D input, or

> *largerRoot* $:= \dfrac{\left(-b + \text{sqrt}\left(b^2 - 4 \cdot a \cdot c\right)\right)}{(2 \cdot a)}$ *# 2-D input*

> $\dfrac{1}{2}\dfrac{-b + \sqrt{b^2 - 4ac}}{a}$

Here are the larger roots of each of the two equations:

> **eval(largerRoot, {a = 2, b = 5, c = −6});**
>
> $-\dfrac{5}{4} + \dfrac{1}{4}\sqrt{73}$

> **eval(largerRoot, {a = 2, b = 1, c = −3});**
>
> $-\dfrac{1}{4} + \dfrac{1}{4}\sqrt{25}$

The above answer can actually be simplified to 1. (Maple does not notice that $\sqrt{25}$ = 5.) You can tell Maple to do so by using the **simplify** command, which will be discussed in detail in the next chapter.

> **simplify(%);**
>
> 1

Functions

**Defining
Functions**

Maple has many built-in functions such as **sqrt**, **sin**, and **tan**. You can define your own functions as well.

To define a function $f(x)$ in Maple, you use:

> **f := x -> *formula* ;**

The "arrow symbol" **->** is formed by entering the minus sign and greater than sign together, with no space(s) in between.

For example, you can define the function $f(x) = x^2 + 5x$ in Maple with:

> **f := x -> x^2+5*x;**   #1-D input, or

> $f := x \rightarrow x^2 + 5 \cdot x$ *#2-D input*

$$x \rightarrow x^2 + 5x$$

Now, we can do some evaluations:

> **f(7.1);**

$$85.91$$

> **f(a);**

$$a^2 + 5a$$

> **f(x+1);**

$$(x+1)^2 + 5x + 5$$

> **f(f(y));**

$$(y^2 + 5y)^2 + 5y^2 + 25y$$

**Functions with
More Than
One Variable**

Functions may have more than one variable. A simple example is the computation of the average speed of an automobile.

If an automobile travels m miles in the span of t minutes, then its average speed in miles per hour is given by the expression $\dfrac{m}{(t/60)} = \dfrac{60m}{t}$. We then have a speed

function " $f(m, t) = \dfrac{60m}{t}$ ":

> **speed := (m,t) -> 60*m/t ;**   #1-D input, or

> $speed := (m, t) \rightarrow \dfrac{60 \cdot m}{t}$ *#2-D input*

$$(m,t) \rightarrow \frac{60m}{t}$$

If a distance of 45 miles is traveled in 30 minutes, the average speed will be 90 m.p.h.:

speed(45,30);

$$90$$

More Examples

Functions of Split Definition

Functions sometimes cannot be defined using a single formula. For example, the famous Heaviside function is defined by:

$$H(x) = \begin{cases} 1, & \text{if } x > 0 \\ 0, & \text{if } x \leq 0 \end{cases}$$

To define this function in Maple, we use the **piecewise** command as follows:

h:= x -> piecewise(x > 0, 1, x <= 0, 0);   #1-D input, or

$h := x \rightarrow piecewise(x > 0, 1, x \leq 0, 0)$ **#2-D input**

$$x \rightarrow \text{piecewise}(x > 0, 1, x \leq 0, 0)$$

(The characters "**<=**" in the command mean $\leq$, less than or equal to.)

To use **piecewise** to define a function having two branches, write:

piecewise(*condition1* , *result1* , *condition2* , *result2*);

This means that if *condition1* is satisfied, then *result1* will be used; otherwise, Maple will use *result2* if *condition2* is satisfied. The following table shows some operators you will use to check conditions.

Operator	Meaning
=	Equal to
>	Greater than
>=	Greater than or equal to
and	And

Operator	Meaning
<>	Not equal to
<	Less than
<=	Less than or equal to
or	Or

The syntax for the **piecewise** command can also be extended. For example, the following function *f* has three branches and can be defined as you see below:

$$f(x) = \begin{cases} 1-x, & \text{if } 1 < x < 3 \\ x^2, & \text{if } 0 \leq x \leq 1 \\ x+2, & \text{if } x < 0 \text{ or } x \geq 3 \end{cases}$$

f := x -> piecewise(1 < x and x < 3, 1-x,
**   0 <= x and x <= 1, x^2, x < 0 or x >= 3, x+2);**

To see this definition more clearly, we recommend that you write the conditions one on each line (using $\boxed{shift - return}$ on some systems) and line them up carefully:

```
f := x -> piecewise( 1 < x and x < 3,    1-x,
                     0 <= x and x <= 1,  x^2,
                     x < 0 or x >= 3,    x+2);
```

Useful Tips

Never assign values to any of the names **x**, **y**, **z**, or **t**. Otherwise, Maple will confuse them with the variables **x**, **y**, **z**, or **t** that you typically use when defining functions.

You should always use **unassign** before defining functions. It will help you to avoid potential conflicts between variable names and function definitions.

Most of Maple's built-in names use only lower case letters. Several names introduced in later versions begin with capital letters. One way to easily distinguish your names from Maple's is to use names that begin with a lowercase letter and include one or more uppercase letters (e.g., **myFunction**, **newVar**).

Even if you start a new worksheet or document in a *single* Maple session, all the variables or functions that you defined in earlier worksheets will still be retained until you quit Maple.

You can define **e := exp(1);**. This will help to simplify many of your inputs. For example, $e^{0.3}$ can now be entered as **e^0.3**, instead of **exp(1)^0.3**. However, it is still preferred to use **exp(0.3)**.

Troubleshooting Q & A

Question... When I defined a new variable or function, I got the error message "`attempting to assign to ... which is protected.`" What did I do wrong?

Answer... If you try to rename a built-in function or built-in constant, Maple will give you this error message. You cannot choose names that conflict with Maple's own names.

Question... I tried to define a function, but I couldn't get it to work. What should I check?

Answer... Three things usually bother function definitions.

- First, check for the proper syntax. Make sure you use a colon-equal definition. Also, don't leave any space between the — and **>** keys in forming the **—>** sign. (Common mistakes include using **=** instead of **:=** and using **f(x):=** instead of **f := x —>**.)

- Second, check that the formula of the function is entered correctly. Some common mistakes are to use expressions such as **cos x**, **e^x**, and so on.

- Third, your function or variable name may conflict with something else you used earlier in your Maple session. For example, you may have once defined:

```
x := 3;
```

Later you define:

```
f := x -> x^2;
f(x);
            9
```

The result is not the x^2 you expected, but $3^2 = 9$ since **x** has the value 3. You can check if this is the case by evaluating **x**. Make sure you explicitly clear the variable(s) before you define your function:

```
unassign('x');
f := x -> x^2;
```

Question...   What is the difference between = (equal sign) and := (colon-equal)?

Answer...   The symbol = means equality, while := means assignment. Equality is a test that gives a true or false answer. Assignment is an action that either gives a name to a value or defines a function.

Question...   When I tried to evaluate a function, I got the error message "**... uses a 2nd argument, ... , which is missing.**" What went wrong?

Answer...   This message means that you did not supply enough variables to evaluate the function.

A common mistake is to define a function of one variable, say,

```
f := x -> x^2;
```

Later, you use the same name **f** to define a function of two variables, say,

```
f := (x, y) -> x + 2*y;
```

The original definition of $f(x) = x^2$ has now been erased. If you type **f(x)**, you will get an error message complaining about not having enough variables (because Maple is now expecting two variables for the function).

Question...   When I used **unassign**, I got the error message "**... cannot unassign.**" What went wrong?

Answer...   Make sure that you enclose the variable name that you want to clear in single quotes. For example:

```
a := 3;
unassign(a);        # This will give you an error message.
Error, (in unassign) cannot unassign `3'
unassign('a');      # This is correct.
```

CHAPTER 5
Computer Algebra

Working with Polynomials and Powers

The expand and factor Commands

In this chapter you will learn how to use Maple to perform common algebraic operations. Let's start with polynomials.

The **expand** command does exactly what its name says it does:

$$\texttt{expand((x-2)*(x-3)*(x+1)\^{}2);} \qquad \text{\# for 1-D input, or}$$

$$expand\big((x-2)\cdot(x-3)\cdot(x+1)^2\big) \qquad \textit{\# for 2-D input}$$

$$x^4 - 3x^3 - 3x^2 + 7x + 6$$

The **factor** command is basically the inverse of the **expand** command:

$$\texttt{factor(x\^{}4-3*x\^{}3-3*x\^{}2+7*x+6);} \qquad \text{\# for 1-D input, or}$$

$$factor\big(x^4 - 3\cdot x^3 - 3\cdot x^2 + 7\cdot x + 6\big) \qquad \textit{\# for 2-D input}$$

$$(x-2)\,(x-3)\,(x+1)^2$$

Here are some more examples:

Example	Comment
`factor(x^2-3);` $$x^2 - 3$$	Although $x^2 - 3 = (x+\sqrt{3})(x-\sqrt{3})$, **factor** will not give radicals in its answer.
`factor(x^2-3,sqrt(3));` $$-(x+\sqrt{3})\,(-x+\sqrt{3})$$	We can give Maple a list of radicals it is allowed to use in factorization.
`expand((x-1.54)*(3.2*x-2.9));` $$3.2x^2 - 7.828x + 4.466$$ `factor(%);` $$3.2\,(x - .90625000)\,(x - 1.540000)$$	**factor** does a nice job even when you use numerical coefficients.
`factor((-2+6*I)+(-5-3*I)*x +` `   (1-I)*x^2);` $$(1-I)\,(-x+1+2I)\,(-x+2I)$$	**factor** works even when the coefficients are complex.
`factor(x^2+1);` $$x^2 + 1$$	**factor** won't use complex numbers unless at least one of the coefficients is a complex number.

`factor(x^2+1, I);` $-(x+\text{I})(-x+\text{I})$	Or you give **factor** explicit permission to use I in its answer.

`expand((x-y+z)^3);`
 	
$x^3 - 3x^2y + 3x^2z + 3xy^2 - 6xyz +$
$3xz^2 - y^3 + 3y^2z - 3yz^2 + z^3$
 	
`factor(%)`
 	
$(x-y+z)^3$
	The **expand** and **factor** commands can also be used for polynomials with more than one variable.

The simplify Command

The **simplify** command tries to produce an expression that Maple thinks is the simplest form. For example:

`simplify(4^(1/2)+5);`   # 1-D input, or,

$$simplify\left(4^{\left(\frac{1}{2}\right)}+5\right)$$ # 2-D input

$$7$$

`simplify((x+2)^2-4);`

$$x^2 + 4x$$

Please note that the **simplify** command will not factor a polynomial into its "simplest" form. That is the job for **factor**.

`factor((x+2)^2-4);`

$$x(x+4)$$

The symbolic Option

By default, Maple is very careful in simplifying. It understands that **sqrt(x^2)** is not always the same as **x** for all complex numbers. Instead, **sqrt(x^2)** simplifies to **x** times the complex sign of **x**.

`simplify(sqrt(x^2));`

$$csgn(x)\,x$$ # csgn stands for complex sign.

Sometimes we want such expressions of powers simplified anyway. In those cases we use the **symbolic** option of the **simplify** command.

`simplify(sqrt(x^2), symbolic);`

$$x$$

The **symbolic** option of the **simplify** command handles $\sqrt{x^2}$ by treating it formally as $(x^2)^{1/2}$ and rewriting it to be $x^{2(1/2)} = x^1$ (without regard for the sign of x). Similarly:

`simplify((x^6)^(1/3));`

$$\left(x^6\right)^{1/3}$$

`simplify((x^6)^(1/3), symbolic);`

$$x^2$$

Working with Rational Functions

The simplify and convert Commands

A rational function is an expression of the form $\dfrac{a\ polynomial}{another\ polynomial}$. The following are three common algebraic operations involving rational functions.

- Combining terms over a common denominator can be done with the **simplify** command. For example, to combine $\dfrac{2}{3x+1}+\dfrac{5x}{x+2}$:

    `simplify( 2/(3*x+1) + (5*x)/(x+2));`

    $$\frac{7x+4+15x^2}{(3x+1)\,(x+2)}$$

- Splitting up rational functions into partial fractions can be done with the **convert** command if you specify the **parfrac** option. For example, to split up $\dfrac{11x^2-17x}{(x-1)^2(2x+1)}$:

    `convert((11*x^2-17*x)/((x-1)^2*(2*x+1)),parfrac,x);`

    $$\frac{5}{2x+1}+\frac{3}{x-1}-\frac{2}{(x-1)^2}$$

- The **convert** command also does long division when you use the **parfrac** option. For example, to find $(x^5-2x^2+6x+1)\div(x^2+x+1)$:

    `convert((x^5-2*x^2+6*x+1)/(x^2+x+1),parfrac,x);`

    $$x^3-x^2-1+\frac{2+7x}{x^2+x+1}$$

Working with Trigonometric and Hyperbolic Functions

Basic Trigonometric and Hyperbolic Identities

The **expand** command can work on expressions that involve trigonometric and hyperbolic functions.

`expand(sin(2*x));`

$$2\sin(x)\cos(x)$$

`expand(cosh(2*x));`

$$2\cosh(x)^2-1$$

The **simplify** command can also be used for trigonometric and hyperbolic functions, but the results may not come out the way you think they should. For example, **simplify** recognizes that $\sin^2 x + \cos^2 x = 1$, but does not recognize the identity $\cos^2 x - \sin^2 x = \cos(2x)$.

> **simplify(sin(x)^2+cos(x)^2);**
>
> $$1$$
>
> **simplify(cos(x)^2-sin(x)^2);**
>
> $$2\cos(x)^2 - 1$$

(Notice that Maple always "simplifies" trigonometric functions by using the cosine function instead of the sine function.)

The combine Command

The **combine** command is more effective in working with trigonometric and hyperbolic functions. For example,

> **combine(cos(x)^2-sin(x)^2);**
>
> $$\cos(2x)$$
>
> **combine(2*sin(x)*cos(x));**
>
> $$\sin(2x)$$
>
> **combine(cosh(x)^2+sinh(x)^2);**
>
> $$\cosh(2x)$$
>
> **combine(sin(2*x)*cos(3*x));**
>
> $$\frac{1}{2}\sin(5x) - \frac{1}{2}\sin(x)$$

Useful Tips

💡 💡 You have to be careful when you use the **symbolic** option in the **simplify** command. Most of the time you use it, you will be assuming that all the quantities you're working with are inside certain subsets of the real numbers.

💡 💡 The **combine** command can also be used very effectively to simplify exponential, logarithmic, and power functions.

Troubleshooting Q & A

Question... When I used **simplify**, **expand**, or **factor** on a polynomial, I got a number. What went wrong?

Answer... First, it is possible that after expansion or simplification, all the terms in your polynomial canceled out and left you with a constant.

If this is not the case, check whether you assigned a value to the variable at some earlier time. For example, you may have assigned

```
x := 5;
```

earlier, then after awhile you typed:

```
expand((x-3)^2*(4*x+5));
```
$$100$$

You asked Maple to expand (5-3)^2*(4*5+5) = 100 which is a number! To correct this, type:

```
unassign('x');
```

and re-execute the **expand** command.

Question... When I used **simplify** or **expand** on a simple polynomial, I got the wrong answer. What went wrong?

Answer... Check to see that you remembered to put an asterisk (*****) between terms being multiplied together. Parenthesized expressions don't get multiplied when they're written next to each other. Consider:

```
expand((x+1)(x+2)^2);
```
$$x(x+2)^2 + 2x(x+2)+1$$

Maple interprets the expression **(x+1)(x+2)^2** as the function $(x+1)^2$ evaluated at $(x+2)$. This is almost certainly not what you wanted to do.

Question... I tried **simplify**, **expand**, **factor**, **combine**, and several other commands on an expression, and I can't get the type of expression I was expecting. What should I do?

Answer... There are more advanced techniques for controlling the way Maple simplifies an expression. But the practical answer may be that the software just can't get you to where you want to be using algebra alone. You may need to look for a different approach.

For example, Maple can't directly simplify $\tanh^{-1}\left((e^x - e^{-x})/(e^x + e^{-x})\right) = x$. But by graphing $\tanh^{-1}\left((e^x - e^{-x})/(e^x + e^{-x})\right)$, you see that it looks like $y = x$. (Chapter 9 does graphing.) You can also compute that $\tanh^{-1}\left((e^x - e^{-x})/(e^x + e^{-x})\right)$ has derivative 1, so that's almost enough to establish the identity. (Chapter 12 will show you how to do derivatives.)

CHAPTER 6

Working with Equations

Equations and Their Solutions

The solve Command for a Single Equation

Maple's **solve** command will solve an equation for an "unknown" variable. You use it in the form:

> **solve(** *an equation* **,** *unknown variable* **);**

For example,

> **solve(2*x+5 = 9, x);**   #1-D input, or,
>
> *solve*$(2 \cdot x + 5 = 9, x)$ **#2-D input**
> $$2$$

Notice that equations in Maple are written using the equal sign "**=**".

You can check that $x = 2$ is the correct solution to the above equation, by evaluating the equation with $x = 2$:

> **eval(2*x+5=9, x=2);**
> $$9 = 9$$

You can even have Maple check that this substitution gives the correct answer by evaluating the result as a Boolean expression.

> **evalb(eval(2*x+5 = 9, x=2));**
> *true*

This means that after the substitution $x = 2$, it is true that the left-hand side of the equation equals the right-hand side.

Here are a few more examples involving **solve**:

Equation	To Solve It in Maple	Comment
Solve $x^2 - 3x + 1 = 0$ for x	**solve(x^2–3*x+1 = 0, x);** $$\frac{3}{2} + \frac{1}{2}\sqrt{5}, \quad \frac{3}{2} - \frac{1}{2}\sqrt{5}$$ **evalf(%);** $$2.618033988, 0.381966012$$	Equations can have more than one solution. We can see a numerical answer with the **evalf** command.
Solve $y^2 - ay = 2a$ for y.	**solve(y^2 – a*y = 2*a, y);** $$\frac{1}{2}a + \frac{1}{2}\sqrt{a^2 + 8a}, \quad \frac{1}{2}a - \frac{1}{2}\sqrt{a^2 + 8a}$$	If the equation involves other variables, Maple will treat them as parameters.

Solve $x^3 + x^2 = -3x$ for x	`solve( x^3+x^2 = -3*x, x );`
	
	$$0, -\frac{1}{2} + \frac{1}{2}I\sqrt{11}, -\frac{1}{2} - \frac{1}{2}I\sqrt{11}$$
	
	`evalf( %, 5 );`
	$$0., -0.50000 + 1.6583I, -0.50000 - 1.6583I$$
		Here two of the solutions are complex. I stands for $\sqrt{-1}$. Numerical approximations up to 5 digits are obtained in this way.

Solve $x + \sin x = \cos x$ for x	`solve( x+sin(x) = cos(x), x );`
	`Warning, solutions may have been`
	`lost`
		No solution is returned when Maple cannot solve an equation.

> **Note:** The **solve** command works very well for equations involving polynomials. However, it often has much less success with trigonometric, exponential, logarithmic, or hyperbolic functions.

The solve Command for a System of Equations

The **solve** command can also be used to solve a system of equations. For two equations in two unknown variables, you use this form of the command:

> **solve({** *equation1*, *equation2* **}, {** *variable1*, *variable2* **});**

■ **Example.** To solve the equations $3x + 8y = 5$ and $5x + 2y = 7$ in the variables x and y, use:

> **solve({3*x+8*y = 5, 5*x+2*y = 7}, {x,y});**
>
> $$\left\{ x = \frac{23}{17}, \ y = \frac{2}{17} \right\}$$

■ **Example.** To solve the equations $3xy - y^2 = -4$ and $2x + y = 3$:

> **solve({3*x*y - y^2 = -4, 2*x + y = 3}, {x,y});**
>
> $$\left\{ \begin{array}{l} x = -\frac{1}{2} RootOf(5\_Z^2 - 9\_Z - 8, label = \_L1) + \frac{3}{2}, \\ y = RootOf(5\_Z^2 - 9\_Z - 8, label = \_L1) \end{array} \right\}$$

Maple gives the answer in terms of the roots of the polynomial $5z^2 - 9z - 8$. (The expression *label* = $\_L1$ indicates that the values of x and y are built from the same root of the polynomial in the answer.) Since the polynomial is quadratic, we expect two sets of solutions. To get a list of solutions, we use the **allvalues** command:

> **allvalues(%);**
>
> $$\left\{ x = \frac{21}{20} - \frac{1}{20}\sqrt{241}, \ y = \frac{9}{10} + \frac{1}{10}\sqrt{241} \right\}, \left\{ x = \frac{21}{20} + \frac{1}{20}\sqrt{241}, \ y = \frac{9}{10} - \frac{1}{10}\sqrt{241} \right\}$$

■ **Example.** If you try

> **solve({x+y = 0, x+y = 1},{x,y});**

No solutions are returned because there are no solutions to this system of equations.

■ **Example.** We can also solve systems of equations involving more than two variables. For example:

```
allvalues( solve( {x+2*y-z=1, x-y+z^2=2, y+x=2*z},
                   {x,y,z} ) );
```

$$\{x = 6\sqrt{2} - 7,\ y = -2\sqrt{2} + 3,\ z = -2 + 2\sqrt{2}\},$$
$$\{x = -7 - 6\sqrt{2},\ y = 3 + 2\sqrt{2},\ z = -2 - 2\sqrt{2}\}$$

Numerical Solutions for Equations

Numerical Answers from solve

The **solve** command can also use efficient numerical techniques to approximate roots of polynomial and a few other simple functions. In order to do this, at least one of the coefficients in the equation has to be a floating point number. For example:

```
solve( x^5-x^3 = 1, x );
```

$$RootOf(\_Z^5 - \_Z^3 - 1, index = 1),\quad RootOf(\_Z^5 - \_Z^3 - 1, index = 2),$$
$$RootOf(\_Z^5 - \_Z^3 - 1, index = 3),\quad RootOf(\_Z^5 - \_Z^3 - 1, index = 4),$$
$$RootOf(\_Z^5 - \_Z^3 - 1, index = 5)$$

Now, if we replace "1" with "1.0" in the equation (i.e., replacing the exact value 1 with its numerical equivalent 1.0), then Maple will report numerical approximations for the answers:

```
solve( x^5-x^3 = 1.0, x );   # use 1.0 instead of 1
```

1.236505703, 0.3407948662 + 0.7854231030 I, −0.9590477179 +0.4283659562 I, −0.9590477179 − 0.4283659562 I, 0.3407948662 − 0.7854231030 I

This trick will also work for systems of equations that involve polynomials or simple functions.

```
solve( {3*x*y - y^2 = 4, 2*x^2 + y = 9.0}, {x,y} );
                                    # use 9.0 instead of 9
```

$$\{x = 1.614445936, y = 3.787128638\},\ \{x = 2.031216860, y = .7483161321\},$$
$$\{x = -2.199468588, y = -0.6753241393\},\ \{x = -2.946194209, y = - 8.360120631\}$$

If you used the same **solve** command without a decimal point in the equation:

```
solve( {3*x*y - y^2 = 4, 2*x^2 + y = 9}, {x,y} );
```

the solution would involve the roots of a quartic equation. Using **allvalues** to obtain explicit representations of the solutions produces several pages of output!

Numerical Solutions Using fsolve

To find numerical solutions for equations in general, use the **fsolve** command. It has the same syntax as **solve**. For a single equation, use:

fsolve(*an equation* , *variable to solve for* **);**

and, for a system of equations, use:

```
fsolve({ equation1, equation2}, { variable1, variable2 });
```

For example:

```
fsolve( x^5-x^3-1=0, x );
```

$$1.236505703$$

In general the **fsolve** command gives a single answer. However, if the equation involves only polynomials of one variable, you can indicate the number of roots to look for with the **maxsols** option. Specify the **complex** option if you want complex values returned as well.

```
fsolve(x^5-x^3-1=0,x, complex, maxsols=5);
```

$-0.9590477179 - 0.4283659563 \, I, \quad -0.9590477179 + 0.4283659563 \, I,$
$0.3407948662 - 0.7854231030 \, I, \quad 0.3407948662 + 0.7854231030 \, I, \quad 1.236505703$

You can usually use **fsolve** to find solutions of a complicated equation whenever **solve** fails.

```
solve( exp(x^2) -50*x^2+3*x = 0, x );
```

   Warning, solutions may have been lost

```
solve( exp(x^2) -50*x^2+3*x = 0.0, x );
```

   Warning, solutions may have been lost

```
fsolve( exp(x^2) -50*x^2+3*x = 0, x );
```

$$-0.115494211$$

The two **solve** commands return no output other than a warning that solutions may be lost; the **fsolve** command returns an approximate numerical solution.

Guiding fsolve with a Range

Using fsolve to Find a Solution in a Specific Interval

Sometimes we are interested in a root other than the one that **fsolve** returns. We can guide **fsolve** by giving it a range in which to look for the solution. The form of the command becomes:

```
fsolve( equation , variable , range for the solution );
```

■ **Example.** We are interested in using **fsolve** to find solutions to the trigonometric equation $\tan(x) = x$. If we know that there is a solution in each of the intervals $-\pi/2 \le x \le \pi/2$, $\pi/2 \le x \le 3\pi/2$, and $3\pi/2 \le x \le 5\pi/2$, we can specify appropriate ranges. We then find three solutions.

```
fsolve( tan(x) -x = 0, x,  -Pi/2.. Pi/2);
fsolve( tan(x) -x = 0, x,  Pi/2..3*Pi/2);
fsolve( tan(x) -x = 0, x, 3*Pi/2..5*Pi/2);
```

$$0$$
$$4.493409458$$
$$7.725251837$$

fsolve with a Range in a System of Equations

To solve a system of two equations for the two unknowns x and y, restricted to the intervals $x_0 \le x \le x_1$ and $y_0 \le y \le y_1$, you use **fsolve** with three arguments:

```
fsolve( { equation1, equation2 }, {x,y},
            { x = x_0..x_1, y = y_0..y_1 });
```

■ **Example.** The system $y^2 - x^3 = 5$ and $y = x - 3\cos x + 4$ has a solution inside the intervals $-2 < x < -1$ and $1 < y < 2$. We can pinpoint it with:

```
fsolve( { y^2-x^3=5, y=x-3*cos(x)+4 }, {x,y},
            { x=-2..-1, y=1..2 } );
```

$$\{ x = -1.231437723,\ y = 1.769915245 \}$$

More Examples

Extracting Solutions from the Results of solve

Sometimes you will want to work with the answers you get from **solve** and **fsolve** without having to retype them. This example will show you the two steps that will enable you to do so.

Consider the equation $x^2 - 3x + 1 = 0$. It has two solutions:

```
solve(x^2-3*x+1 = 0, x);
```

$$\frac{3}{2} + \frac{1}{2}\sqrt{5}, \quad \frac{3}{2} - \frac{1}{2}\sqrt{5}$$

• Step 1. Make the output of the **solve** command the contents of a list and give the list a name. (We discuss lists in the next chapter.)

```
ans := [ solve(x^2-3*x+1 = 0, x) ];
```

$$\left[\frac{3}{2} + \frac{1}{2}\sqrt{5}, \quad \frac{3}{2} - \frac{1}{2}\sqrt{5} \right]$$

• Step 2. Identify each of the answers with **ans[1]** and **ans[2]**.

```
ans[1];
```

$$\frac{3}{2} + \frac{1}{2}\sqrt{5}$$

```
ans[2];
```

$$\frac{3}{2} - \frac{1}{2}\sqrt{5}$$

Let's check that **ans[2]** is really a solution to $x^2 - 3x + 1 = 0$:

```
simplify( eval(x^2 - 3*x + 1, x=ans[2]) );
```

$$0$$

How about:

```
(x - ans[1])*(x - ans[2]);
```

$$\left(x - \frac{3}{2} - \frac{1}{2}\sqrt{5}\right)\left(x - \frac{3}{2} + \frac{1}{2}\sqrt{5}\right)$$

```
expand( % );
```

$$x^2 - 3x + 1$$

Useful Tips

 fsolve works much more quickly using *numeric methods* than **solve** does using *algebraic methods*.

fsolve can deal with all kinds of equations, but it only gives a single answer in most cases. (It can be given a range in which to look for a solution, but that removes the automation.) **solve** will try to find up to 100 solutions, but in many cases, cannot solve the equations symbolically.

We think that the best approach is first to use **solve**, including a floating point number in one of the equations to find a numerical solution. Then, work with the output!

Maple can return all solutions to some transcendental equations when the environment variable **_EnvAllSolutions** is set to true. For example:

```
solve(sin(x) = 1/2);    # The answer is in the first quadrant.
```

$$\frac{1}{6}\pi$$

```
_EnvAllSolutions := true:
solve(sin(x) = 1/2);    # The general solution.
```

$$\frac{1}{6}\pi + \frac{2}{3}\pi\_B1 \sim +2\pi\_Z1 \sim$$

(In the output, _B1 ~ is a binary constant (0 or 1) and _Z1 ~ is an integer constant.)

The following table can help you remember the difference between the various types of brackets. But beware – they are not interchangeable!

Syntax Element	Purpose	Example
(*parentheses*)	(i) Grouping terms in a computation	`(x^2+3)*(x-1)`
	(ii) Arguments of commands	`sin(x^2)`
[*square brackets*]	List (i.e. an ordered list)	`[x, y, z]`
{ *curly braces* }	Set (i.e., an unordered list)	`{x-3 = y,` `5*x+y = 2}`
List [*square brackets*]	Specify a position in a list	`soln[1]`

Troubleshooting Q & A

Question... When I used **solve** or **fsolve**, I got an error message. What should I check first?

Answer... You should first check if you made a mistake in the input of the equation(s).

- Check that you entered the formula correctly. Common mistakes include misspelling names of the Maple built-in functions, and forgetting to type the multiplication symbol "*****".

- Did you remember to include the variable(s) to solve for in the input?

- If you are using more than one equation, make sure all the curly braces and commas are located in the right places.

Question... When I used **solve** or **fsolve**, I got the error message "a constant is invalid as a variable." What should I check?

Answer... Check that the variable or variables you are trying to solve for have no assigned values. For example, you may have earlier assigned:

```
x := 3;
```

Later, you try:

```
solve( x^2-1=0, x);
Error, (in solve) a constant is invalid as a variable, 3
```

You should have cleared the variable(s) before using **solve**.

```
unassign('x');
```

Question... I used **solve** and Maple gave me the output "*RootOf*". What does this mean?

Answer... For example, if you try:

```
sol := [solve(sqrt(x) = (x-3)^2, x)];
```

$$[\,RootOf(-\_Z + \_Z^4 - 6\_Z^2 + 9,\ index = 1)^2,$$
$$RootOf(-\_Z + \_Z^4 - 6\_Z^2 + 9,\ index = 2)^2\,]$$

Maple tells you that the solutions to the equation $\sqrt{x} = (x-3)^2$ are the squares of the 1st and 2nd roots of the polynomial $-z + z^4 - 6z^2 + 9$. This is not helpful. However, you can see the numerical value of these answers with:

```
evalf({sol[1], sol[2]});
```

$$\{1.835964860, 4.452626878\}$$

If the equation involves polynomial only, you can also use the **allvalues** command to convert the answer to a friendlier form:

```
allvalues(%);
```

Question... I used `solve` but could not understand Maple's output. It had the symbols %1, %2, and so on. What does this mean?

Answer... If the result from `solve` is too lengthy, Maple will use the symbols %1, %2, etc., to represent expansions of subexpressions that occur several times.

Question... Maple did not give any output from `solve` or `fsolve`. Why not?

Answer... This usually means one of three things.

- Maple does not know how to solve your equation(s) with the `solve` command.

- If you used `fsolve` with a specified interval, Maple was unable to locate a root in that interval. Make sure that the interval(s) you gave contains a root.

- There is no solution to your equation or system of equations. A single equation might reduce to an absurdity (e.g., $x = x + 1$ reduces to $0 = 1$). A system of equations may be inconsistent (e.g., $x + y = 1$ and $x + y = 2$).

Question... When I used `fsolve`, I got an error message that says "`...should use exactly all the indeterminates.`" or "`...is in the equation, and is not solved for.`" What happened?

Answer... *Maple* cannot use a numerical method to approximate the solution, because your equation involves a constant that is not defined. For example, if **a** is not defined, *Maple* can do:

```
solve(x^2 = a, x);
```

But you will get an error message with:

```
fsolve(x^2 = a, x);
```

Question... `fsolve` failed to give me a solution to an equation in an interval where I know there is a solution. What happened?

Answer... `fsolve` uses a version of Newton's method to search for an answer. The procedure will not always find an answer. It is particularly vulnerable to functions whose graphs have narrow spikes on otherwise well-behaved regions.

For example, away from the origin, $\tan(x) = x^3$ has this feature, and it is hard for `fsolve` to locate a root without a very precise range.

Question... After solving an equation, I get real and complex roots. How can I only select the real roots?

Answer... To remove any term from *your list* that contains "`I`", use the following `remove` command.

```
remove(has, your list, I);
```

For example,

```
soln := [solve(x^7-5*x+4=0.0, x)];
```

[1., 0.8851280726, 0.5229014897 + 1.183774264 I, −0.7607643824 + 1.155768213 I,
−1.409402287, −0.7607643824 − 1.155768213 I, 0.5229014897 − 1.183774264 I]

```
realSoln := remove(has, soln, I);
```

[1., 0.8851280726, −1.409402287]

Only the real solutions are selected.

Similarly, if you want to select those roots that have non-zero imaginary part, you can use:

```
imaginarySoln := select(has, soln, I);
```

You will learn more about **select** and **remove** in Chapter 24.

CHAPTER 7
Sets, Lists, and Sequences

Lists and Sets

What Is a List?   A **list** in Maple is an expression in which elements are separated by commas and enclosed in **[** *square brackets* **]**. For example, each of the following is a list.

```
[2, 5, 7, 10, -3, -25];          # Each element is a number.
```
[2, 5, 7, 10, -3, -25]

```
[3, [1,2], 6, `good`, `bad`, -2, `ugly`];
```
                    #The elements consist of numbers, a list of numbers and names.

[3, [1,2], 6, *good, bad*, -2, *ugly*]

What Is a Set?   A **set** in Maple is an expression in which elements are separated by commas and enclosed in **{** *curly braces* **}**. For example, each of the following is a set.

```
{2, 5, 7, 10, -3, -25};          # Each element is a number.
```
{-25, -3, 10, 7, 5, 2}

```
{3, [1,2], 6, `good`, `bad`, -2, `ugly`};
```
                    #The elements consist of numbers, a list of numbers and names.

{–2, 3, 6, *bad, good, ugly*, [1, 2]}

The Difference between Sets and Lists   A **list** is an ordered set. We can have repetition in the elements. On the other hand, ordering and redundancy of elements in a **set** does not matter.

```
{5,4,3,2,1};      # The arrangement of the elements in a set doesn't matter.
```
{1, 2, 3, 4, 5}

Also, duplicate elements are automatically removed from a set, but not from a list:

```
{1,2,2,3,3,3,4,4,4,4,5,5,5,5,5};
```
{1, 2, 3, 4, 5}

```
[1,2,2,3,3,3,4,4,4,4,5,5,5,5,5];
```
[1, 2, 2, 3, 3, 3, 4, 4, 4, 4, 5, 5, 5, 5, 5]

Lists and sets are important structures in Maple. Many of Maple's inputs and outputs are expressed using lists or sets. For example, Maple gives us two sets for the output when we solve this system of equations:

solve({x+y=2, x^2+y = 2}, {x,y});

$$\{x = 1, y = 1\}, \ \{x = 0, y = 2\}$$

We can then collect the solutions to form a list:

ans := [solve({x+y=2, x^2+y = 2},{x,y})];

$$[\{x = 1, y = 1\}, \{y = 2, x = 0\}]$$

Working with Elements of a List

Maple lets you work with the elements of a list directly. You can access each element using the **[** *square brackets* **]** notation.

For example, consider the list,

myList := ["red", 2, 3, 3.153, [1,3,5], sin(z), Pi, "history", Fred];

$$["red", 2, 3, 3.153, \ [1, 3, 5], \sin(z), \ \pi, "history", Fred]$$

The following table shows some examples and how you can work with the elements of this list:

Maple Command	Explanation
myList[4]; 3.153	Returns the 4th element of the list.
myList[-3]; π	Returns the entry that is 3rd from the end of the list.
myList[3..6]; [3, 3.153, [1,3,5], sin(z)]	Returns a list of the 3rd through 6th elements of the **myList**.

op(myList[3..6]);
    3, 3.153, [1,3,5], sin(z)
or you can use
myList[3..6][];
    3, 3.153, [1,3,5], sin(z)
	The **op** command will let us see the elements without a list.
	
	The empty **[** *square brackets* **]** at the end will give the same result.

myList[5];
    [1,3,5]
myList[5][2];
    3
myList[5,2];   #same as above
    3
	If an element of a list is a list itself, we can retrieve its entries in the same manner, using **[** *square bracket***]** syntax.

myList[2] := 88;
    88
myList;
    ["red", 88, 3, 3.153, [1, 3, 5], sin(z), π, "history", Fred]
	This syntax lets you directly change the 2nd element of **myList**.
	
	You can see that 2nd element of this list has been replaced.

Sequences

The seq Command

The **seq** command can be used to generate a sequence of elements that can be defined by a mathematical formula. For example, to make a sequence of the form n^2, for each integer $1 \le n \le 16$, we will type:

```
seq(n^2, n=1..16);
```

$$1, 4, 9, 16, 25, 36, 49, 64, 81, 100, 121, 144, 169, 196, 225, 256$$

In general, the **seq** command is used in the form:

seq(*expression in a variable n*, $n = n_0 .. n_1$ **);**

Here are a few examples:

Maple Command	Remark
`seq( n^2, n=-2..8 );` $4, 1, 0, 1, 4, 9, 16, 25, 36, 49, 64$	The variable n runs from -2 to 8.
`seq( [cos(n),n/(n+1)], n = 1..3 );` $\left[\cos(1), \dfrac{1}{2}\right], \left[\cos(2), \dfrac{2}{3}\right], \left[\cos(3), \dfrac{3}{4}\right]$	The expression in the **seq** command can be a list itself. In this case, we form a table of coordinate pairs.
`seq( x^n, n=0..7 );` $1, x, x^2, x^3, x^4, x^5, x^6, x^7$	Each element can be a symbolic expression.

By combining the **seq** command with **[** *square brackets* **]** or **{** *curly braces* **}**, we can create a list or set. For example,

```
[ seq( n, n=1..22 ) ];
```

$$[1, 2, 3, 4, 5, 6, 7, 8, 9, 10, 11, 12, 13, 14, 15, 16, 17, 18, 19, 20, 21, 22]$$

```
{ seq( cos(n), n =1..5 ) };
```

$$\{\cos(1), \cos(2), \cos(3), \cos(4), \cos(5)\}$$

More Examples

An Experiment in Factoring

Let us look at the factorization of $x^n + 1$. With the help of the **seq** command, we can see the factorization for several values of n very quickly, say for $n = 2, 3, \ldots, 10$.

```
seq(factor(x^n+1), n=2..10);
```

$$x^2 + 1, \ (x+1)(x^2 - x + 1), \ x^4 + 1, \ (x+1)(x^4 - x^3 + x^2 - x + 1),$$
$$(x^2 + 1)(x^4 - x^2 + 1), \ (x+1)(1 - x + x^2 - x^3 + x^4 - x^5 + x^6), \ x^8 + 1,$$
$$(x+1)(x^2 - x + 1)(x^6 - x^3 + 1), \ (x^2 + 1)(x^8 - x^6 + x^4 - x^2 + 1)$$

Do you notice that all the coefficients in all the factors above are either 1 or –1? Try repeating the command with **n = 2..100**. You will see the same thing!

You may conclude that every coefficient in every factor of $x^n + 1$ will be ±1, no matter what n is. But before you celebrate your latest discovery, check the factorization of $x^{105} + 1$. Surprise!

Drawing a Nonagon

■ **Example**. A regular nonagon is a regular polygon of nine sides (i.e., all nine sides have the same length and all nine angles are equal). It can be formed by joining together the points with coordinates $\left(\cos(2\pi n/9), \sin(2\pi n/9)\right)$, where $n = 1, 2, \ldots, 9$.

We can create a list of these points using the **seq** command:

```
coordlist:=
    [seq([cos(n*2*Pi/9),sin(n*2*Pi/9)],n=0..9)];
```

$$\Big[[1, 0], [\cos(\tfrac{2}{9}\pi), \sin(\tfrac{2}{9}\pi)], [\cos(\tfrac{4}{9}\pi), \sin(\tfrac{4}{9}\pi)], [-\tfrac{1}{2}, \tfrac{1}{2}\sqrt{3}],$$
$$[-\cos(\tfrac{1}{9}\pi), \sin(\tfrac{1}{9}\pi)], [-\cos(\tfrac{1}{9}\pi), -\sin(\tfrac{1}{9}\pi)], [-\tfrac{1}{2}, -\tfrac{1}{2}\sqrt{3}],$$
$$[\cos(\tfrac{4}{9}\pi), -\sin(\tfrac{4}{9}\pi)], [\cos(\tfrac{2}{9}\pi), -\sin(\tfrac{2}{9}\pi)], [1,0]\Big]$$

If we join the points one by one, we will create the nonagon. This can be done using the **plot** command. (**plot** will be discussed in details in Chapters 9 and 10.)

```
plot( coordlist );
```

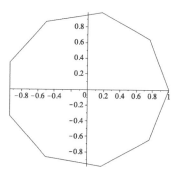

Troubleshooting Q & A

Question... When I used the **seq** command, Maple gave me no output. What does that mean?

Answer... This means that the sequence you asked for is empty, because the index range you specified does not make sense. Make sure that $n_0 \le n_1$ in the command

```
seq(expression, n=n₀..n₁);
```

CHAPTER 8

Getting Help and Loading Packages

Getting Help

Help for Specific Maple Commands

You can always get some help on how to use a *Maple* command by typing the command, highlighting it, going to the **Help** menu in the menu bar, then selecting "**Help on** *command*" - the fifth item in the menu. A help window will open, giving you details of the command, including options and examples.

For example, if you forget how to use the **fsolve** command, you can type **fsolve,** highlight this string, then select "**Help on fsolve**" from the **Help** menu.

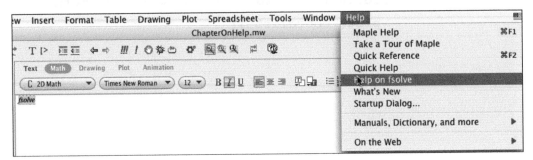

The help window that comes up gives help for the command and a list of other related commands.

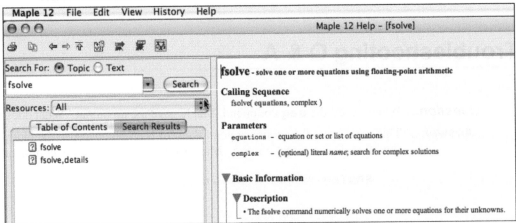

Looking Up a Command

Sometimes, you won't remember the exact name of a command that you want use. Say, you remember that it started with **plo**, but you just can't recall its full name. You can type **plo**, highlight it, and select "**Help on plo**" from the **Help** menu. The help browser will open with Maple's best guess as to the command you're looking for. It will also list many other options for you on the left side in case it guessed wrong.

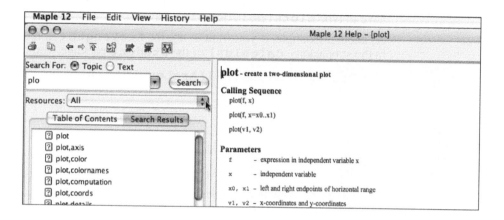

Packages

Loading a Package

Maple has a built-in vocabulary of several hundred commands. But still these are not enough for everyday usage. Additional commands are available in the Maple "packages." A few of the more common packages are: **Algebraic**, **Calculus**, **Linear Algebra**, **plots**, **Statistics**, **Student**, etc..

Before using a command defined in a package, you have to load the package using the **with** command. For example, the **animate** command is defined in the **plots** package. We will load the package with:

```
with(plots);
```

[animate, animate3d, animatecurve, arrow, changecoords, complexplot, complexplot3d, conformal, conformal3d, contourplot, contourplot3d, coordplot, coordplot3d, densityplot, display, dualaxisplot, fieldplot, fieldplot3d, gradplot, gradplot3d, graphplot3d, implicitplot, implicitplot3d, inequal, interactive, interactiveparams, intersectplot, listcontplot, listcontplot3d, listdensityplot, listplot, listplot3d, loglogplot, logplot, matrixplot, multiple, odeplot, pareto, plotcompare, pointplot, pointplot3d, polarplot, polygonplot, polygonplot3d, polyhedra_supported, polyhedraplot, rootlocus, semilogplot, setcolors, setoptions, setoptions3d, spacecurve, sparsematrixplot, surfdata, textplot, textplot3d, tubeplot]

The output shows all of the commands defined in the **plots** package that have been loaded. (You can hide this list by using a colon to terminate the previous command line.)

Now we can use the **animate** command:

```
animate(sin(2*x*t), x=-2..2, t=-2..2);
```

Now, click on the picture, then click the play button ▶ near the top of the window and the show starts! (Chapter 26 discusses animation in detail.)

Using a Command Without Loading the Package

You can also load and use a single command from a package without loading the whole package. For example, if you only want to use the **polygonplot** command defined in the **plots** package, you can type:

    plots[polygonplot](*input values for the command* **);**

Maple will then know where to find the definition of that command. However, since you have not loaded the **polygonplot** command, the next time you use it, you have to type **plots[polygonplot]** again.

As another example, if you want to use the **ArcLength** command which is defined in the subpackage **Calculus1** of the **Student** package, you type:

    Student[Calculus1][ArcLength] (*input values for the command***);**

Useful Tips

The help pages contain examples showing how the commands can be used. Pick the one or two sample commands that are closest to what you are trying to do, execute them, and see how they work.

Another way of invoking help is to use the **help** or **?** commands:

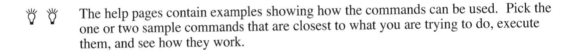

    help(*command* **);**   #or you can use

    ?*command***;**

For example, to get help on the **fsolve** command, you can type:

    ?fsolve |*enter*|

When you load a package, finish the command with a colon instead of a semicolon,

    with(*package* **):**

Then Maple will not display the whole list of the commands in that package, which can be very lengthy.

You can also load a package from the menu bar. In the **Tools** menu, select "**Load Package**", then select the package you want to load. The list shows the most commonly-used packages, and the last item "**List All Packages...**" lets you see the complete list of packages available.

To unload a package, use the **unwith** command.

    unwith(*package* **):**

CHAPTER 9

Making 2-D Pictures

Drawing the Graph of a Function

The plot Command

A common operation in mathematics is to plot the graph of a function, such as a polynomial, over a given interval. Maple does this with its **plot** command. To plot a function of x over an interval $a \le x \le b$, you type:

```
plot( function, x = a..b );
```

For example, the following plots the graph of $y = x^3 - 3x^2 + 5x - 10$ for x in the interval $-2 \le x \le 3$.

```
plot(x^3-3*x^2+5*x-10, x=-2..3);
```

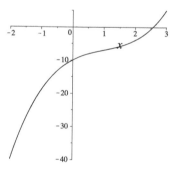

Plot Options Tool Bar

When you click once on a picture from the **plot** command, the **Plot options** tool bar will show up near the top of the worksheet. By clicking various buttons in the options bar, you can alter the appearance of the picture with different plot styles, axes styles, and scales.

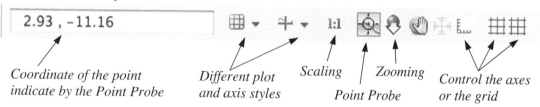

Coordinate of the point indicate by the Point Probe

Different plot and axis styles

Scaling

Zooming

Point Probe

Control the axes or the grid

For example, the following graph shows the result of plotting the same graph as above, then choosing the boxed icon for the axes and adding a grid of lines.

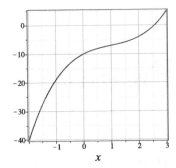

**Plot Options
Menus**

When you click on a picture, the sub-menus of the **Plot** menu in the menu bar at the top of the Maple window become active, as you see in the picture below. This lets you control even more options for the plot.

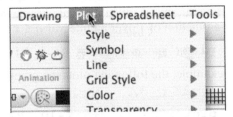

You can then ask Maple to redraw the graphic by choosing various options from these menus. For example, we can repeat the **plot** command above, then add a title, and legend, change the style to points, and make the symbol a cross to obtain the graph below.

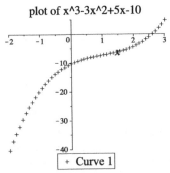

The following table summarizes some of the feature of a graph that can be controlled using the submenus of the Plot menu.

Menu	Comment
Style	Choose if graphs are drawn as curves or collections of points
Symbol / Line	Choose styles for lines and points in a picture.
Legend	Add and edit legends to a picture. Most useful when the plot has more than one function.
Title	Add and edit a title and caption to the picture

Axes	Choose a style for the axes. The **properties** option opens a dialog box that lets you set the viewing window for the plot.
Projection	Choose either **constrained** or **unconstrained** scaling. Serves same purpose as the 1:1 button in context menu bar.
Manipulator	Controls the action produced by a mouse click or drag in the plot.
Export	Lets you export the picture in various formats, such as postscript, GIF, JPG, and so on.

> **Note**: You should use **constrained** scaling in the **Projection** menu or the 1:1 button on the **Plot options** tool bar, whenever you want to see angles or circles properly. For example, a circle can look like an ellipse unless you specify **constrained** scaling.

More Options with the plot Command

Plotting Multiple Functions

The **plot** command allows you to graph several functions or expressions simultaneously, all on the same set of axes, over a common interval $a \le x \le b$. You use the following format.

```
plot([ function1, function2, etc. ], x = a..b );
```

For example, to see the graphs:

$$y = 3x^4 - 5x, \quad y = 10\sin(x) - 10, \text{ and } y = 5\cos(x) + 3e^x$$

on the interval $-2 \le x \le 2$, you type:

```
plot([3*x^4-5*x, 10*sin(x)-10, 5*cos(x) + 3*exp(x)],
     x = -2..2);
```

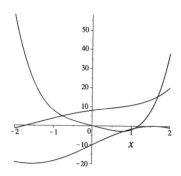

Notice the use of [*square brackets*] to group the functions together. This indicates the order in which the graphs are going to be added to the plot. A drawback to this command is that all the functions are plotted on the same domain (in this case **x =**

-2..2). Later in this chapter, we will see another method to combine different pictures.

When drawing multiple functions, you may also want to use a legend to identify the graphs. You do this by choosing the **Show Legend** option in the **Legend** menu under the **Plot** menu. You can also use the **Edit Legend** menu to specify a legend.

Restricting the y-axis

If a graph contains a vertical asymptote or has a very large range of y-values, some of its interesting features may not be visible. That's because Maple tries to show you the whole graph. The solution is to tell Maple what portion of the viewing window you want to see.

For example, the graph of the tangent function has vertical asymptotes. Consider the difference between Maple's default picture and one where we restrict the graph to show only y-values with $-10 \le y \le 10$.

```
plot( tan(x), x=-6..6 );
```

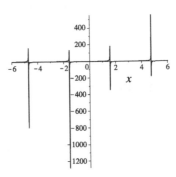

```
plot( tan(x), x=-6..6, y=-10..10 );
```

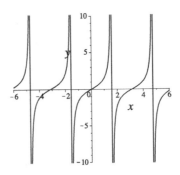

Working with Asymptotes and Discontinuities

Maple draws asymptotes in the picture above because the curve is constructed by connecting points, left to right, within the specified domain. You can instruct Maple to not connect points across a discontinuity by adding the optional argument **discont=true** in the **plot** command.

```
plot( tan(x), x=-6..6 , y=-10..10, discont=true);
```

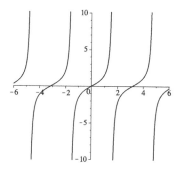

Labeling Pictures

You can add a title to a plot and change the labels on the axes by using the **title** and **labels** options in the **plot** command.

```
plot( 15+cos(x), x=0..4*Pi ,
    labels = [ "day", "price" ],
    title = "Daily price of Stock ABC" );
```

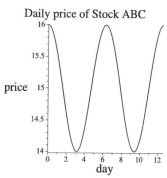

More Advanced Drawings

Colors

You can change the color of a curve by specifying the **color** option inside the **plot** command. You have 155 pre-defined colors to choose from. Some of the colors Maple knows are:

```
"Aquamarine", "Black", "Blue", "Navy", "Coral",
"Cyan", "Brown", "Gold", "Green", "Grey", "Magenta",
"Maroon", "Orange", "Pink", "Plum", "Red", "Tan",
"Turquoise", "Violet", "Wheat", "White", and "Yellow".
```

Say, let us draw a picture in orange:

```
plot(x^3+2*x, x=-3..2, color = "Orange");
```

Sorry, we cannot show you the picture here, because this text is printed in black and white!

Alternatively, we could use the **COLOR** function with specified **RGB** values to create a color. This is done using a **COLOR** structure in which the amounts of red, blue, and green light in the final color are specified. For example, **COLOR(RGB,1,0,0)** is red, while **COLOR(RGB,1,0,1)** is purple.

We can redraw the same picture in orange using **RGB** values of 1, 0.647 and 0.

```
plot(x^3+2*x, x=-3..2, color = COLOR(RGB,1,0.647,0));
```

Thickness

One way to distinguish curves in a black-and-white graph is to control the thickness of a curve. This is done by setting the **thickness** option to have the value 1 for a thin pen (default), 2 for a medium pen, 3 for a thick pen, 4 for a thicker pen, etc.. The larger the value, the thicker will be the graph.

```
plot(x^3+2*x, x=-3..2, thickness = 3);
```

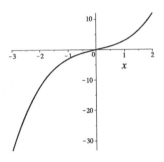

Combining Plots with the display Command

You can combine different plots with different styles into a single picture by using the **display** command, which is loaded as part of the **plots** package.

First, name each of the plots. Then, use the **display** command to combine these named graphics into a single output.

```
graph1 := plot( cos(x), x=-5..1, color=green):
graph2 := plot( sin(x), x=0..6,
                      thickness=2, color=yellow):
graph3 := plot( x^2-1, x=-1.5..1.5,
                      thickness=3, color=red):
```

We ended each command above with a colon instead of a semicolon. The colon serves the same purpose as a semicolon, except that the colon indicates that Maple should perform the desired calculation internally without displaying the result.

In this case, if we had used a semicolon, we could have still stored the plots, but then Maple would have displayed a harmless output "*PLOT(...)*."

```
with(plots):    # We have to load the plots package to use display.
display( [ graph1, graph2, graph3 ] );
```

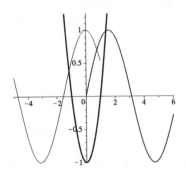

Annotating a Plot with Text and Drawings

The Drawing Tool Bar

Maple lets you annotate a plot by adding text and drawings. When you click once on a plot, the **Plot** options tool bar will show up near the top of the worksheet. Click on the **Drawing** button (see the picture below), the Drawing tool bar will show up with the standard drawing tools. You can use these tools to work directly on the plot.

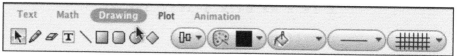

For example, we can start with the **plot** command:

```
plot(sin(x)+.2*x, x = -10 .. 10);
```

The **Drawing** tool can be used to indicate the local maximum and minimum points of the graph. This gives the annotated graph.

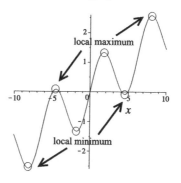

More Examples

Zooming In

■ **Example.** Consider $f(x) = \sqrt{1 + 10x^4 - 20x^5 + 25x^6}$ and $g(x) = x^2 + 5\sin(x)$:

```
f := x -> sqrt( 1 + 10*x^4 - 20*x^5 + 25*x^6);
g := x -> x^2 + 5*sin(x);
plot([f(x),g(x)], x=-3..3);
```

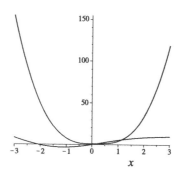

To get a better estimate of the intersection near 1, we can zoom in on the graph by successively shrinking down the *x*-interval in the **plot** command.

```
plot([f(x),g(x)], x=1 .. 1.5);
plot([f(x),g(x)], x=1.13 .. 1.15);
plot([f(x),g(x)], x=1.135 .. 1.145);
```

"Optical Illusion"

■ **Example.** Suppose we plot the function $f(x) = 64x^4 - 16x^3 + x^2$:

```
f := x -> 64*x^4-16*x^3+x^2;

plot(f(x), x=0..10);
```

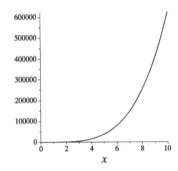

You can easily get the impression that the graph is always increasing. However, notice that the vertical range is very large, between 0 and 600,000, so the picture is not sharp enough to show any small dips. In particular, the graph looks like a straight line for *x* between 0 and 2; this certainly is not the case.

```
plot(f(x), x=0..2);      # Still looks OK.
plot(f(x), x=0..1);      # Still looks OK.
plot(f(x), x=0..0.5);    # Still looks OK.
plot(f(x), x=0..0.2);    # Surprise!!
```

Useful Tips

☼ ☼ ☼ If you're trying to arrange several graphs with different plot options on the same graph, it's always easier to draw the graphs separately and then use the **display** command to put them together.

☼ ☼ ☼ Maple uses a variety of colors to draw 2-D graphs. These may look pleasing on screen, but may not look very good when printed on a gray scale printer. We suggest that unless you have a color printer available, you add the option **color = black** to your plots.

☼ ☼ If you have a **plot** that involves complicated functions, always try it first with one function at a time to make sure that each **plot** comes out OK by itself.

You will find other methods of making 2-D pictures in Chapters 10 and 11. Also, there are many other options in **plot** that we have not discussed. You can find them using **?plot[options]** and experiment with them.

Troubleshooting Q & A

Question... I got an empty picture from **plot** with an error message about "...empty plot." What went wrong?

Answer... This usually indicates that Maple cannot evaluate your input function numerically. Check whether you made a typo in the input. Some common mistakes are:

- You typed the name of a built-in function incorrectly.

- You used the wrong variable.

- You specified an interval in which the input function is not defined.

Check whether your function really gives numbers! Do you get numbers when you enter **f(-1.)**, **f(0.)**, **f(1.)** and so on?

Question... When I combined various pictures using the **display** command, I got the reply *display(PLOT(......* but no picture. Where was my mistake?

Answer... Make sure that you load the **plots** package before using the **display** command. Type:

```
with(plots);
```

and re-enter your **display** command.

Question... When I combined various plots using the **display** command, I got an error message. Where was my mistake?

Answer... Check if you typed the names of the plots correctly in your **display** command. If there is no misspelling there, then you should **display** each picture one at a time in order to find out which one is causing the problem. Then recheck their definitions. (A common mistake is to use = instead of := in defining the plots.)

CHAPTER 10

Plotting Parametric Curves Line Segments, and Points

Parametric Curves

Plotting Parametric Curves

In the previous chapter you saw how to use **plot** to draw curves that are graphs of functions. But not all curves are the graphs of functions.

A two-dimensional (2-D) parametric curve is written in the form $(x(t), y(t))$. The **plot** command can be used to draw the curve $(x(t), y(t))$, defined on an interval $a \leq t \leq b$. The command has the form:

> **plot([x(t), y(t), t = a..b]);**

For example, to see the curve $(t^2 - 1, t + t^2)$ for $-2 \leq t \leq 2$, we use:

> **plot([t^2-1, t+t^2, t=-2..2]);**

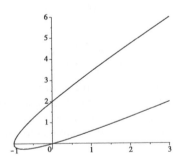

Plot Options

The options that we discussed in the previous chapter (e.g., **color, scaling** and **thickness**) also work for this particular format of **plot** command. For example, to draw the unit circle $(\cos t, \sin t)$ for $0 \leq t \leq 2\pi$ with a blue color, and the two axes measured in the same scale, you use:

> **plot([cos(t),sin(t), t=0..2*Pi], thickness = 3,**
> **scaling = constrained, color = blue);**

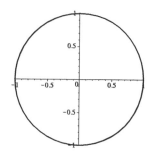

Plotting Line Segments and Polygonal Paths

Line Segments   The **plot** command can also be used to plot a line segment joining the points (x_0, y_0) and (x_1, y_1). The syntax is:

> **plot([[x_0, y_0], [x_1, y_1]], x = $a..b$);**

For example, to see the line segment joining (0, 1) to (–1, 3):

> **plot([[0,1],[-1,3]], x=-2..2);**

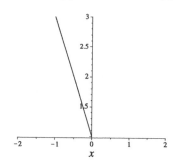

Polygonal Paths   A similar construction is used to plot a polygonal path connecting n points.

> **plot([[x_0, y_0], [x_1, y_1], ... ,[x_n, y_n]], x = $a..b$);**

For example, to plot the triangle with vertices (0, 1), (1, 3), and (2, –1) you type:

> **plot([[0,1],[1,3],[2,-1],[0,1]], x=-1..3);**

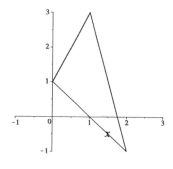

Plotting Individual Points

To have Maple plot only points without drawing the line segments that connect them, include the optional argument **style=point**. The shape and size of each point is controlled by the **symbol** and **symbolsize** options, respectively. This is demonstrated in the following example.

```
plot([[0,1],[1,3],[2,-1], [0.5,1.5]], style = point,
      symbol = solidcircle, symbolsize = 18);
```

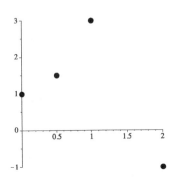

Plotting Multiple Curves in One Plot

Multiple Curves at Once

You can plot several parametric curves, graphs, and lines all in one picture, using a single **plot** command. To do this, you put all the expressions of the curves inside [*square brackets*].

■ **Example.** To draw the curves (t^2, t) for $-2 \leq t \leq 2$, and (t, t^2) for $-1 \leq t \leq 1$, the line segment from (1, 0) to (2, 3), and the graphs $y = 2x$, $y = x^2$ for $-2 \leq x \leq 3$, we can use:

```
plot( [ [t^2,t, t=-2..2], [t,t^2, t=-1..1],
        [[1,0],[2,3]], 2*x, x^2 ], x=-2..3 );
```

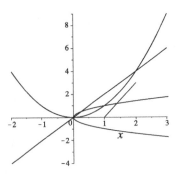

Multiple Pictures Together

You can also combine several 2-D pictures that you've already created with the **display** command. (The **display** command is defined inside the **plots** package. This approach is used when the pictures you want to combine do not have the same viewing window or do not all use the same options.

■ **Example**. To create the "open skull" picture that you see below, we will combine several pictures. The face is made from a parabola and two straight lines; the eyes from two circles; and the mouth from a straight line.

```
face := plot([x^2, 4, 4.5+x/4],x=-2..2, color = red):
eyes := plot([ [1+cos(t)/4, 3+sin(t)/4, t=0..2*Pi],
               [-1+cos(t)/2, 3+sin(t)/2, t=0..2*Pi]],
               color = brown):
pupil := plot([[-1,3],[1,3]], style=point,
               color=green, symbol=solidcircle,
               symbolsize=18 ):
mouth := plot([[-0.5,1],[0.5,1]], color = blue):
with(plots):
display([face,eyes,pupil,mouth],
        scaling=constrained, axes=none );
```

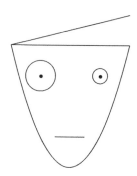

Useful Tips

☿ ☿ ☿   You may be overwhelmed by all the different plotting formats that we have discussed. The following table can help you to remember them:

Expression of the Form	*What It Plots*
`plot( [` $f_1(x)$`,` $f_2(x)$ `],` **x** `=` $a..b$`);`	Graphs of $f_1(x)$, $f_2(x)$ for $0 \leq x \leq b$
`plot( [` $x(t)$`,` $y(t)$`,` **t** `=` $a..b$`] );`	Parametric curve $(x(t), y(t))$, $a \leq t \leq b$
`plot( [[`a`,` b`], [`c`,` d`]] );`	Line segment from (a, b) to (c, d)
`plot([[`a`,` b`], [`c`,` d`]],` `    style=point);`	Points (a, b) and (c, d)

☿ ☿   If you need to draw a complicated picture that consists of several graphs or curves, always draw each picture individually and combine them with the **display** command. You may be able to use a single **plot** command to draw the entire picture, but if you get an error message, it can be difficult to locate the mistakes.

Troubleshooting Q & A

Question... I tried to draw one parametric curve with **plot**, but instead Maple gave me a picture of two curves. What went wrong?

Answer... When entering a parametric curve, it is common to forget to include the interval *inside* the **[** *square brackets* **]**. For example, instead of typing **plot([t, t^2, t = −2..2]);** you may make the mistake of typing:

plot([t, t^2], t=-2..2);

Then Maple will draw the *graphs* $y = t$ and $y = t^2$!! This is not what you want.

Question... I tried to draw a parametric curve with **plot** but got an error message "Error, (in plot) unexpected options: ...." What should I check?

Answer... This error message suggests that you should check the syntax of the **plot** command. Make sure that you used **[** *square brackets* **]** and **(** *parentheses* **)** correctly . For example, a common mistake is to input the expression **(t, t^2)** or **(t, t^2, t = −2..2)** instead of **[t, t^2, t = −2..2]** for the parametric curve $(t, t^2), -2 \le t \le 2$.

Question... I tried to draw a parametric curve with **plot** but got an error message "Error, (in plot) expecting a real constant as range ...." What should I check?

Answer... This error message suggests that you made a mistake in specifying the range for the parameter. For example, check to see if you used a variable to define the range. Also, a common mistake is to use **pi** instead of **Pi**.

Question... I tried to graph a parametric curve with **plot** but got an error message. What went wrong?

Answer... This usually indicates that Maple cannot evaluate your input function numerically. Check whether you made a typo in the input. Common mistakes include:

- You typed the wrong name of a built-in function.

- You used the wrong variable.

- You specified an interval in which the input function is not defined.

Question... I tried to use one of the commands from the **plots** package but Maple just echoes back the command. Why won't Maple execute the command?

Answer... The most common cause of this problem is not loading the **plots** package. Enter and execute the command **with(plots):** and re-execute your command.

CHAPTER 11

Polar and Implicit Plots

Plotting in Polar Coordinates

The polarplot Command

Curves expressed in polar coordinates can be graphed with the **polarplot** command. To access this command it is necessary to load the **plots** package. (See the discussion of packages in Chapter 8.)

```
with(plots):
```

If the curve is given by $r = f(\theta)$, for $\theta_1 \le \theta \le \theta_2$, we can draw it with the following:

```
polarplot (f(θ), theta = θ₁ .. θ₂);
```

Notice that this form is very similar to that of the **plot** command you already know. For example, to plot the three-leaf rose, use:

```
with(plots):
polarplot( 2*cos(3*theta) , theta = 0..2*Pi );
```

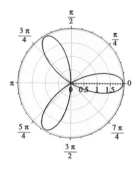

Plotting Multiple Curves with Style

Just as with the **plot** command, several curves can be plotted at once with the **polarplot** command by enclosing them in [*square brackets*], separated by commas. Also, you can supply many of the options that work with the **plot** command.

■ **Example.** A command to graph the spirals $r = \dfrac{\theta}{2\pi}$, $r = \left(\dfrac{\theta}{2\pi}\right)^2$, and the circle $r = 1$, all in one plot, is:

```
polarplot( [ theta/(2*Pi), (theta/(2*Pi))^2, 1 ],
          theta=0..2*Pi, axes=none,
          linestyle=[1,3,6], color=black);
```

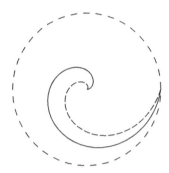

Notice how the **linestyle** option is used to make the curves easily identifiable in a black-and-white plot by dashing lines. The **axes=none** instructs Maple to omit the polar axes.

Cartesian Plotting Tricks Revisited

When plotting in polar coordinates, we face many of the same issues we found in Cartesian coordinates in order to get a nice picture. For example, the graph of $r = \tan(\theta)$ has discontinuities and produces very large values of r. Thus the command:

```
polarplot(tan(theta), theta=0..2*Pi);
```

will not show much detail, especially near the origin. We will get a better picture by limiting the values of r and θ using the **coordinateview** option. Also we use **discont=true** to take care of the spurious lines from the discontinuities.

```
polarplot( tan(theta), theta=0..2*Pi,
          coordinateview=[0..5, 0..2*Pi],
          discont=true);
```

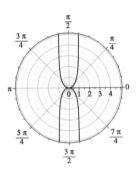

Plotting Graphs of Equations

The implicitplot Command

When a curve is given by an equation in variables x and y, you can draw the curve with the **implicitplot** command (defined in the **plots** package).

You use the command in the form:

```
with(plots):
implicitplot( an equation in x and y , x = a..b, y = c..d);
```

For example, to see the unit circle $x^2 + y^2 = 1$ for $-1 \le x \le 1$, $-1 \le y \le 1$, use:

```
with(plots):
implicitplot( x^2 + y^2 = 1, x=-1..1, y=-1..1);
```

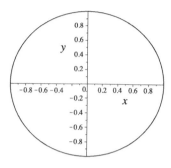

(Note that the circle is not round because the x and y axes are not of the same scale. You can correct this by adding the option **scaling = constrained**.)

Multiple Curves and Styles

You can also use [*square brackets*] or the **display** command to plot multiple implicit curves. The format is the same as you've used already with the **plot** and **polarplot** commands. Many of the options you use with the **plot** command work with **implicitplot** as well.

■ **Example.** We want to see the curves $x^2 + 3xy + y^3 = 25$, $x^2 + 3xy + y^3 = 10$, and $x^2 + 3xy + y^3 = 0$ in the intervals $-10 \le x \le 10$ and $-6 \le y \le 4$. Let's draw each curve with a different color:

```
f := (x,y) -> x^2+3*x*y+y^3;
pict1 := implicitplot( f(x,y) = 25, x=-10..10,
            y=-6..4, color = red):
pict2 := implicitplot( f(x,y) = 10, x=-10..10,
            y=-6..4, color = yellow):
pict3 := implicitplot( f(x,y) = 0, x=-10..10,
            y=-6..4, color = green):
display( [pict1,pict2,pict3] );
```

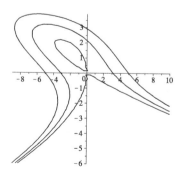

This picture looks like the Chinese character for "Wind." We can get the same plot with a single call to **implicitplot** as follows:

```
implicitplot( [ f(x, y)=25, f(x, y)=10, f(x, y)=0 ],
            x = -10..10, y = -6..4,
            color = [red, yellow, green]);
```

Useful Tips

☿ ☿   You may be overwhelmed by all the different plotting commands that we have discussed. Don't worry, because **plot** is the most commonly used command, while **polarplot** and **implicitplot** are for special situations:

To Draw	Use
Graph of a function $f(x)$	**plot**
Parametric curve $(x(t), y(t))$	**plot**
Curve in polar coordinates	**polarplot**
Curve given by an equation	**implicitplot**

☿ ☿   Beginning with Maple 12, the **polarplot** command adds a polar grid and uses constrained scaling by default. The grid can be removed either by modifying the plot with the toolbar after execution or by using the **axes=none** or **gridlines =false** option in the command. You can change the axes to Cartesian with the **axiscoordinates = cartesian** option.

☿ ☿   The **display** command lets you combine the results from different kinds of 2D-plot commands (such as **plot**, **polarplot**, **implicitplot**) on a single graph.

☿ ☿   A parametric curve of the form $x = r(t)\cos(\theta(t)),\ y = r(t)\sin(\theta(t))$, can be plotted using polar coordinates with the **polarplot** command or the **plot** command (using the **coords=polar** option) in the following formats.

```
polarplot([r(t), θ(t),t = a..b]);
plot([r(t), θ(t),t = a..b], coords=polar);
```

☿   The **algcurves** package can be used for graphing polynomial equations implicitly. The **plot_real_curve** command in this package usually draws a more accurate curve. For example:

```
implicitplot(x^2=y^3, x=-3..3, y=-3..3,
             view=[-3..3,-3..3]);
```

and

```
with( algcurves ):
plot_real_curve( x^2-y^3, x, y, view=[-3..3,-3..3] );
```

draw the same curve, but the **plot_real_curve** command works better near the critical point.

Troubleshooting Q & A

Question... When I tried to use the **polarplot** or **implicitplot** command, Maple gave the command back without doing anything. What happened?

Answer... Usually, there are two reasons why this might happen:

- You probably forgot to load the **plots** package before using these commands. Enter the following, and then try your commend again.

    ```
    with(plots):
    ```

- Check to see that you correctly spelled the command names **polarplot** or **implicitplot**.

Question... I got a warning message "**... unable to evaluate the function**" and then Maple showed a picture of a horizontal line with labeled polar coordinate axes. Is that picture correct?

Answer... No, that picture is not correct. Check whether you made a typo in the input. Common mistakes are:

- You mistyped the name of a built-in function/variable.

- Your input function contains a variable other than **theta** (a common mistake is to include **r**, the radius variable, in the input).

- You did not match up parameters correctly (e.g., you used **t** in the function but used **theta** when you specified the interval).

Question... I got an error message from **implicitplot**. What should I look for?

Answer... There are two major problem areas in using **implicitplot**.

- Make sure that the equation you entered is really an equation with an equal sign "=". Check that the equation is entered correctly. Did you remember to type * for multiplication?

- Make sure that your equation has two variables that do not have values. Executing **unassign('x','y');** before using **x** and **y** in your equation for **implicitplot** is highly recommended.

Question... The plot produced by **implicitplot** is very coarse, or it missed some parts of the level curve. What can I do to improve the quality of the plot?

Answer... The **implicitplot** command works by searching a mesh of points in the viewing window. By default this grid is 26×26. The size of this grid can be controlled by the **grid** option in **implicitplot**. For example,

```
f := (x,y) -> x^2+y^2:
implicitplot( [f(x,y)=1, f(x,y)=0.73, f(x,y)=0.31],
              x=-3..3, y=-3..3, grid=[90,90] );
```

CHAPTER 12

Limits and Derivatives

Limits

The limit Command

If f is a function of a single variable, Maple evaluates $\lim_{x \to a} f(x)$ with the following syntax:

> **`limit( function, x = a );`**

For example, Maple agrees that $\lim_{x \to 0} \dfrac{\sin x}{x} = 1$.

```
limit( sin(x)/x, x = 0 );
              1
```

We can also enter this command using palettes (see Chapter 2) as follows:

Keystrokes: $\boxed{\lim_{x \to a} f}$ **x** *tab* **0** *tab* $\boxed{\sin(a)}$ **x** $\boxed{\rightarrow}$ $\boxed{\rightarrow}$ **/** **x** $\boxed{\rightarrow}$ $\boxed{enter}$

where $\boxed{\lim_{x \to a} f}$ and $\boxed{\sin(a)}$ are selected from the **Expression** palette, *tab* is the *tab key*, and $\boxed{\rightarrow}$ is the right arrow key.

The following table shows some sample limit computations, demonstrating that Maple can compute most limits, even those that involve infinite limits or limits at infinity:

Limit Calculation	*In Maple*
$\lim\limits_{x \to 0} \dfrac{e^x - 1 - x}{x^2} = \dfrac{1}{2}$	`limit((exp(x)-1-x)/x^2, x = 0);` $\dfrac{1}{2}$
$\lim\limits_{x \to 3} \dfrac{1-x}{(x-3)^2} = -\infty$	`limit((1-x)/(x-3)^2, x = 3);` $-\infty$
$\lim\limits_{x \to +\infty} \dfrac{x}{\sqrt{x^2+1}} = +1$ $\lim\limits_{x \to -\infty} \dfrac{x}{\sqrt{x^2+1}} = -1$	`f := x -> x/sqrt(x^2+1);` `limit(f(x),x = infinity);` 1 `limit(f(x),x = -infinity);` -1
$\lim\limits_{x \to 0} \dfrac{\lvert x \rvert}{x}$ does not exist	`limit(abs(x)/x, x=0 );` *undefined*

One-sided Limits

To calculate a left-hand limit $\lim\limits_{x \to a^-} f(x)$, you have to add the **left** option in the **limit** command. To find $\lim\limits_{x \to 0^-} \dfrac{\sqrt{2x^2}}{x}$, you use:

> **limit(sqrt(2*x^2)/x, x = 0, left);**
>
> $$-\sqrt{2}$$

Similarly, to find the right-hand limit, you use:

> **limit(sqrt(2*x^2)/x, x = 0, right);**
>
> $$\sqrt{2}$$

Differentiation

Differentiation Using the diff Command

Maple uses **diff** for computing derivatives. You use it in the form:

> **diff(*function, variable*);**

For example,

> **diff(x^4, x);**
>
> $$4x^3$$

Or you can use $\boxed{\dfrac{\mathrm{d}}{\mathrm{d}x} f}$ from the **Expression** palette.

Keystrokes: $\boxed{\dfrac{\mathrm{d}}{\mathrm{d}x} f}$ **x** *tab* **x^4** $\boxed{\rightarrow}$ $\boxed{enter}$

To calculate a second derivative, you could use **diff** twice:

> **diff(diff(x^4, x), x);**
>
> $$12x^2$$

Or you can use either of these short-hand formats:

> **diff(x^4, x, x);**

> **diff(x^4, x$2);**

Similarly, the third derivative is returned by either of the following:

> **diff(y^4, y, y, y);**

> **diff(y^4, y$3);**
>
> $$24y$$

A few more examples of how you move from mathematical notation to Maple notation for derivatives are given in the following table.

Mathematical Expression	*Maple Evaluation*
$$\dfrac{d(x^2 + e^{x^3})}{dx}$$	`diff( x^2 + exp(x^3), x );` $$2x + 3e^{x^3}x^2$$
$$\dfrac{d(\cos(y^2) + y^5)}{dy}$$	`diff( cos(y^2) + y^5, y );` $$-2\sin(y^2)y + 5y^4$$
$$\left.\dfrac{df}{dt}\right|_{t=1}, \text{ where } f(t) = t^2 + t^3 + \ln(t)$$	`f := t -> t^2 + t^3 + ln(t);` `eval( diff( f(t), t ), t=1);` $$6$$

Differentiation Operator

The last example above shows that the syntax of evaluating derivatives can sometimes be tricky. Maple has a **D** command that can help us. It computes the derivative function. Here's how you'd use it:

```
f := t -> t^2+t^3+ln(t):

df := D(f);                    # defines the function f′
```
$$t \rightarrow 2t + 3t^2 + \frac{1}{t}$$

```
df(t);                         # calculates f′(t)
```
$$2t + 3t^2 + \frac{1}{t}$$

```
df(1);                         # calculates f′(1)
```
$$6$$

> Note: You use **diff** when you want an expression for the derivative, but you use **D** when you want the derivative function.

Differentiation Rules

Maple knows all the formal computation rules in differentiation, such as:

```
unassign( 'f','g' );      # Clears any previous definition of f and g.
diff( f(x)*g(x), x );     # The product rule
```
$$\left(\frac{d}{dx}f(x)\right)g(x) + f(x)\left(\frac{d}{dx}g(x)\right)$$

```
diff( f(x)^n, x );        # The power rule.
```
$$\frac{f(x)^n \, n\left(\frac{d}{dx}f(x)\right)}{f(x)}$$

```
diff( f(g(x)), x );       # The chain rule
```
$$D(f)(g(x))\left(\frac{d}{dx}g(x)\right)$$

How about the rule for differentiating the product of three functions?

```
diff( f(x)*g(x)*h(x), x);
```

$$\left(\frac{d}{dx}f(x)\right)g(x)\,h(x) + f(x)\left(\frac{d}{dx}g(x)\right)h(x) + f(x)\,g(x)\left(\frac{d}{dx}h(x)\right)$$

More Examples

Limits and Graphs

■ **Example.** We compute $\lim\limits_{x\to 0}\dfrac{\sin(\sin(2x)^2)}{x^2}$:

```
limit(sin(sin(2*x)^2)/x^2, x = 0);
```
$$4$$

We can check both graphically and numerically to see if this answer is correct.

- *(Graphically)* Use the graph to determine if the height of the function approaches 4, as x approaches zero.

```
delta := 0.01;
plot( sin(sin(2*x)^2)/x^2, x = 0-delta..0+delta );
```

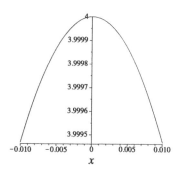

- *(Numerically)* Calculate the values of the function at various test values of x that are close to zero. Check if they approach 4:

```
f := x -> sin(sin(2*x)^2)/x^2;
[f(-0.1), f(0.01), f(-0.001), f(0.0001)];
```
$$[3.945925593, 3.999466587, 3.999994668, 3.999999948]$$

A neater way to show this result is to use the "do loop" that we will discuss in Chapter 27.

```
h := 1.0:
for n from 1 to 4 do
  h := -h/10:
  print('h' = h, 'f(h)' =f(h));
end do:
```

$$h = -0.100000000000, \quad f(h) = 3.945925593$$
$$h = 0.0100000000000, \quad f(h) = 3.999466587$$
$$h = -0.0010000000000, \quad f(h) = 3.999994668$$
$$h = 0.00010000000000, \quad f(h) = 3.999999948$$

The slickest way to construct a list of function values is with the **map** command, discussed in Chapter 24.

```
map( f, [-0.1, 0.01, -0.001, 0.0001] );
```
$$[3.945925593, 3.999466587, 3.999994668, 3.999999948]$$

Definition of Derivative

■ **Example.** Consider the function

```
f := x -> x^2*sin(x) + cos(x):
```

We know that the derivative is $f'(x) = \lim_{h \to 0} (f(x+h) - f(x))/h$. When h is sufficiently small, say $h = 0.1$, we would expect $(f(x+0.1) - f(x))/0.1$ to be very close to $f'(x)$. We can see this by plotting these two expressions on the same graph:

```
plot( [diff(f(x),x), (f(x+0.1)-f(x))/0.1], x=-3..3 );
```

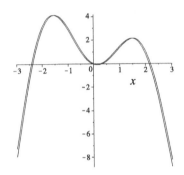

The result can be even better if we choose $h = 0.01$:

```
plot( [diff(f(x),x),(f(x+0.01)-f(x))/0.01], x=-3..3);
```

Geometry of the Derivative

■ **Example.** Consider

```
f := x -> 3*x^2 - 6*x*cos(x):
df := D(f);
```

Then the value of the derivative at $x = a$, $f'(a)$, gives the slope of the tangent line at that point. Using this result, the equation of the tangent line is given by:
$$y = f(a) + f'(a)(x-a).$$

We can check this by combining the graph of $y = f(x)$ with the tangent lines at $x = -2.7$, $x = -1.4$ and $x = 0.9$. (Recall that **D(f)** means f'.)

```
graph1 := plot( f(x), x=-3.5..2, thickness=3 ):

line1 := plot(f(-2.7)+df(-2.7)*(x+2.7), x=-3.5..-2):
line2 := plot(f(-1.4)+df(-1.4)*(x+1.4), x=-2..0):
line3 := plot(f(0.90)+df(0.90)*(x-0.90), x=0..2):

plots[display]( [graph1, line1, line2, line3] );
```

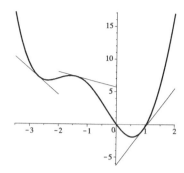

Useful Tips

The commands **limit** and **diff** have inert versions that start with a capital letter (**Limit** and **Diff**). Each returns its argument unevaluated in mathematical notation.

If you're having trouble with a limit, start with the inert version of the command to make sure that you have the expression correct. Then change to lower case and evaluate the command. For example:

```
Diff( x*sin(x)+exp(x^2)+2*x, x ) ;
```

$$\frac{d}{dx}\left(x\sin(x)+e^{(x^2)}+2x \right)$$

```
diff( x*sin(x)+exp(x^2)+2*x, x ) ;
```

$$\sin(x)+x\cos(x)+2x\,e^{(x^2)}+2$$

```
Limit( (exp(x)-1-x)/x^2, x=0 )
  = limit( (exp(x)-1-x)/x^2, x=0 );
```

$$\lim_{x \to 0}\frac{e^x-1-x}{x^2}=\frac{1}{2}$$

Here's another way to obtain the last expression.

```
q := Limit( (exp(x)-1-x)/x, x=0 ):
q = value( q );
```

You can check your syntax on a complicated **limit** or **diff** command by converting the expression to 2-D math input. Highlight the expression and choose **Format** → **Convert to** → **2-D Math Input** from the menu.

You can also use the context menu to convert the expression. Position the cursor over the expression and right-click (control-click on a one-button mouse). The context menu is displayed. Select **Convert to -> 2-D Math Input**.

Inside the **Student[Calculus1]** package, there are commands such as **LimitTutor** and **DiffTutor** that present a convenient graphical interface for students to learn more about a particular Calculus topic. Please see Appendix A for more information about this package.

Troubleshooting Q & A

Question... Maple returns my limit expression unevaluated. What does that mean?

Answer... This means that Maple cannot determine the value of the limit. You may be able to see what the limit is by looking at the graph of the function. (Recheck the first example in the More Examples section.)

Question... I get zero for the derivative of a non-constant function. What should I check?

Answer... You should check the following:

- Are you differentiating with respect to the correct variable? For example, you may have defined a function in terms of t but differentiated with respect to x.

- A common mistake is to type **diff(f, x)** to find the derivative when f is a function. This is wrong, because this command means the differentiation of the *function f* and not the *expression $f(x)$*. You have to type **diff(f(x), x)** or **D(f)**.

Question... I cannot assign a new function which depends on the **diff** command. What should I look for?

Answer... The usual way of defining a new function using : = doesn't work very well with the **diff** command. For example, if you try

```
g := x -> diff(x^2, x) ;
g(x);
```
$$2x$$

It looks OK, but when you ask, say, for the value of g at $x = 2$

```
g(2);
Error, (in g) invalid input: diff received 2, which is not
valid for its 2nd argument
```

This is wrong because Maple is trying to calculate **diff(2^2,2)** which does not make sense. The correct way is to define g using the **D(f)** command. (Please note that the output of the **D** operator is a function.)

```
f := x -> x^2;
g := D(f);
```
$$x \rightarrow 2x$$

```
g(2);
```

4

CHAPTER 13
Integration

Antidifferentiation

The int Command

You can use the **int** command to compute indefinite and definite integrals $\int f(x)\,dx$ and $\int_a^b f(x)dx$. The basic form for an indefinite integral is:

```
int( function, variable );
```

You can also click on $\boxed{\int f\ dx}$ in the **Expression** palette and fill in the template.

For example, to compute $\int x^2\,dx$:

```
int(x^2, x);
```

$$\frac{1}{3}x^3$$

> **Note:** Maple does not automatically include the constant of integration, "+ C", in its antiderivatives.

Maple can integrate almost every integral that can be done using standard integration methods (e.g., substitution, integration by parts, partial fractions), plus many others using special techniques not commonly taught. Here are some typical integrations:

Integral	In Maple	Comment
$\int x\ln(x)\,dx$	`int( x*ln(x), x );` $$\frac{1}{2}x^2\ln(x) - \frac{1}{4}x^2$$	This integral is evaluated using integration by parts.
$\int\dfrac{y^2}{\sqrt{1-y^2}}\,dy$	`int( y^2/sqrt(1-y^2), y );` $$-\frac{1}{2}y\sqrt{1-y^2} + \frac{1}{2}\arcsin(y)$$	This one is found using a trigonometric substitution (try $y = \sin u$).
$\int \sin(\cos(y^2))\,dy$	`int( sin(cos(y^2)), y );` $$\int \sin(\cos(y^2))\,dy$$	This integrand does not have a closed-form antiderivative.
$\int e^{-x^2}\,dx$	`int( exp(-x^2), x );` $$\frac{1}{2}\sqrt{\pi}\ \mathrm{erf}(x)$$	This antiderivative is not an elementary function (see **?erf**).

When Maple can't evaluate an integral, it usually returns your input unevaluated. This can mean either that it's not possible to find an antiderivative in closed form or that Maple hasn't yet been programmed to do the integral. In other cases Maple may return an answer that involves "special functions." Examples include the error function (**erf**) and the sine- and cosine-integrals (**Si** and **Ci**). See the Q & A section for more explanation.

Definite Integration

The int Command, Revisited

A definite integral $\int_a^b f(x)\,dx$ is computed in Maple with this form of the **int** command:

```
int( f(x),   x = a .. b );
```

Or you can use $\left|\int_a^b f\,dx\right|$ from the from the **Expression** palette.

Maple will try to find an antiderivative first, then evaluate it at the endpoints and subtract (according to the FTC – Fundamental Theorem of Calculus). Here are some examples:

Integral	In Maple	Comment
$\int_1^2 x^2\,dx$	`int( x^2, x=1 .. 2 );` $\dfrac{7}{3}$	$\dfrac{x^3}{3}\Big\vert_1^2 = \dfrac{8}{3} - \dfrac{1}{3} = \dfrac{7}{3}$
$\int_2^\infty \dfrac{1}{5+t^2}\,dt$	`int( 1/(5+t^2), t=2 .. infinity );` $-\dfrac{1}{5}\arctan\left(\dfrac{2}{5}\sqrt{5}\right)\sqrt{5} + \dfrac{1}{10}\pi\sqrt{5}$	This improper integral involves $+\infty$.
$\int_0^1 \sqrt{\cos(x^2)}\,dx$	`int( sqrt(cos(x^2)), x=0 .. 1 );` $\int_0^1 \sqrt{\cos(x^2)}\,dx$	When there's no antiderivative, Maple can't apply the FTC.
$\int_0^1 \sin(x^2)\,dx$	`int( sin(-x^2), x = 0 .. 1 );` $-\dfrac{1}{2}\mathrm{FresnelS}\left(\dfrac{\sqrt{2}}{\sqrt{\pi}}\right)\sqrt{2}\,\sqrt{\pi}$	FresnelS (the Fresnel Sine integral) is a special function.

When Maple returns a definite integral unevaluated or in terms of a special function that you do not know, it is usually best to look for a numerical approximation.

Numeric Integration

The evalf and Int Commands

The **Int** command is the inert form of integration. It's like the **Limit** and **Diff** commands seen in the previous chapter. When Maple sees **Int** , it does not attempt to evaluate the integral (definite or indefinite). Used by itself, it's a very useful way to check that you have entered the integrand correctly. Except for the capital "**I**", the syntax is the same as for the **int** command:

> **Int(** *function, variable* **);**
>
> **Int(** *function, variable* = *a* **..** *b* **);**

For example, the following does no evaluation even though we all know it's really easy:

```
Int(x, x=0..1);
```

$$\int_0^1 x \, dx$$

When **evalf** sees an unevaluated definite integral, Maple uses a numerical method to approximate the integral. In other words, we can find a numerical approximation for a definite integral with the following syntax:

> **evalf(Int(** *function, variable* = *a* **..** *b* **));**

This combination (**evalf** and **Int**) can calculate approximations for almost all definite integrals, including $\int_0^1 \sqrt{\cos(x^2)} \, dx$ for which **int** failed (as you saw earlier). Also, the computation is quick.

■ **Example.** To get approximate, numerical values for the integrals $\int_0^2 x^2 dx$ and $\int_0^1 \sqrt{\cos(x^2)} \, dx$, use these commands:

```
evalf( Int( x^2, x = 0..2 ) );
                    2.666666667
evalf( Int( sqrt(cos(x^2)), x = 0..1 ) );
                    0.9485216211
```

More Examples

Exact and Approximate Value of an Integral

■ **Example.** Consider the integral $\int_0^3 \sqrt{9 - x^2} \, dx$. We can see its inert form with:

```
q := Int( sqrt(9-x^2), x=0..3 );
```

$$\int_0^3 \sqrt{9 - x^2} \, dx$$

In the previous chapter, we mentioned that the **value** command could find the value of an inert form:

    value(q);   # The exact value of the integral

$$\frac{9}{4}\pi$$

    evalf(q);   # An approximate value of the integral
                    7.068583471

    evalf(value(q));   #Floating point approximation of exact value
                    7.068583472

The fact that the last two outputs are (slightly) different is an indication that these numbers were obtained by different methods.

Area Between Curves

■ **Example.** To approximate the area bounded by the curves $p(x) = x^5 - 20x^3$ and $q(x) = 30 - x^5$, we start by sketching the curves. First, let's see where they intersect:

    p := x -> x^5 - 20*x^3;
    q := x -> 30 - x^5;

    fsolve(p(x) = q(x), x);
            -3.080038835, -1.206322273, 3.231778500

The values –3.08, –1.206 and 3.232 give approximations for the x-coordinates of the intersections. We can see the area between these two curves in the interval $-3.1 \le x \le 3.3$:

    plot([p(x), q(x)], x = -3.08 .. 3.232);

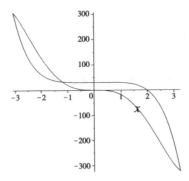

The area can be approximated with $\int_{-3.08}^{3.232} |p(x) - q(x)|\, dx$ using **int**:

    int(abs(p(x)-q(x)), x=-3.08..3.232);
                388.8535522

Fundamental Theorem of Calculus

■ **Example.** The Fundamental Theorem of Calculus states that if f is a continuous function on the interval $[a ,b]$, then $\dfrac{d}{dx}\displaystyle\int_{a}^{x} f(t)\,dt = f(x)$ on the interval $[a, b]$. Maple knows about this theorem.

```
unassign('f');
F := int( f(t), t = a..x);
```

$$\int_{a}^{x} f(t)\,dt$$

```
diff( F, x );
```

$$f(x)$$

Maple can also apply the Fundamental Theorem together with the Chain Rule when both the upper and lower limits involve the variable x.

```
G := int( f(t), t=sqrt(x)..x^2 );
```

$$\int_{\sqrt{x}}^{x^2} f(t)\,dt$$

```
diff( G, x );
```

$$2x\,f(x^2) - \frac{1}{2}\frac{f(\sqrt{x})}{\sqrt{x}}$$

Plotting an Antiderivative

■ **Example.** Maple cannot find an explicit formula for the antiderivative

$$\int 100\ln(x)e^{-x^2}\,dx .$$

```
f := x->100*ln(x)*exp(-x^2):
int(f(x),x);
```

$$\int 100\ln(x)\mathbf{e}^{-x^2}\,dx$$

However, you can see the graph of one of the antiderivatives with the help of the **evalf** and **int** commands. According to the Fundamental Theorem of Calculus, the function $F(x) = \displaystyle\int_{1}^{x} 100\ln(t)\mathbf{e}^{-t^2}\,dt$ is an antiderivative of $100\ln(x)\mathbf{e}^{-x^2}$. Maple can compute values of F using numerical integration (**evalf** and **Int**) and you can then **plot** these values.

```
plot( evalf( Int(f(t), t=1..x) ), x=1..3 );
```

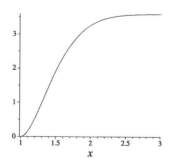

Useful Tips

☿ ☿   Always use **evalf** and **Int** to evaluate a definite integral, unless you need an exact answer. In many cases, **int** will neither work nor give you a useful result. Even when **int** does work, it can be slow.

☿   When working with inert integrals (or limits or derivatives), it is recommended that you use Maple's labels to refer to the inert object. For example, you can type **evalf(** ⎡*control* − ⎤**L**, *enter the label number corresponds to the inert object*, click *OK, and finish the typing with* **)** ;

☿   In Appendix A, we describe the **Student[Calculus1]** package in Maple which includes many commands such as **AntiderivativePlot**, **ApproximateInt**, **AntiderivativeTutor**, **IntTutor**, **SurfaceOfRevolution**, and **VolumeOfRevolutionTutor**, to help students investigate and experiment with integration and some of its applications.

☿   When **int** is used to evaluate a definite integral, if one or both of the limits of integration is a floating-point number such as **0.0**, or **exp(-0.1)** but not **Pi** or **exp(1/10)**, then Maple automatically uses a numerical method to evaluate the definite integral. For example,

```
int( sqrt(cos(x^2)), x=0..1 );        # no result
int( sqrt(cos(x^2)), x=0..1.0 );      # replace 1 by 1.0
```
$$0.9485216211$$

Troubleshooting Q & A

Question... I got some strange answers for definite integrals that used names like "**erf**," "**fresnelS**," and "**fresnelC**." What happened?

Answer... Maple knows about a lot of special functions that appear quite often in integration problems. For example, **erf** is the Maple name for the error function

$$erf(x) = \frac{2}{\sqrt{\pi}} \int_0^x e^{-t^2} dt$$

This is an integral that does not have a closed-form antiderivative, but its values are well known.

You might see something like this:

```
int(exp(-t^2), t=0..1);
```
$$\frac{1}{2}\sqrt{\pi}\, erf(1)$$

```
evalf(%);
```

$$.7468241330$$

Maple has reported $\int_0^1 e^{-t^2} dt = \frac{\sqrt{\pi}}{2}(\frac{2}{\sqrt{\pi}}\int_0^1 e^{-t^2} dt) = (\frac{\sqrt{\pi}}{2})erf(1) \approx 0.747.$

Question... When I evaluated a definite integral with a parameter in the limits of integration, I got a warning message "unable to determine if ... is between ... and ...." What should I do?

Answer... Since a parameter is used in describing the interval of integration, Maple could not determine whether the function was well defined inside this interval. For example,

```
int(1/(x^3+1), x=0..a);
```

```
Warning, unable to determine if -1 is between 0 and a ...
```

$$\int_0^a \frac{1}{x^3+1} dx$$

Maple is unhappy because it knows the function is not continuous on the interval $[0,a]$ unless $a > -1$. Here's what happens when this extra information is provided:

```
int(1/(x^3+1), x=0..a) assuming a > -1;
```

$$\frac{1}{18}\sqrt{3}\,\pi - \frac{1}{6}\ln(a^2+1-a) + \frac{1}{3}\sqrt{3}\arctan\left(\frac{2}{3}\sqrt{3}\,a - \frac{1}{3}\sqrt{3}\right) + \frac{1}{3}\ln(a+1)$$

Question... What is the difference between using "**evalf** and **int**" vs "**evalf** and **Int**"?

Answer... When you use **evalf** and **int**, Maple first tries to find an antiderivative. If one is found, then the definite integral is evaluated by the FTC and then this answer is evaluated numerically. This is not numerical integration.

When you use **evalf** and **Int**, Maple applies a numerical method without spending any time trying to find an antiderivative.

CHAPTER 14
Series and Taylor Series

Series

The add Command

You can use the **add** command to add up a finite number of numeric terms in a sequence. To find $\sum_{n=n_0}^{n_1} (expression\ of\ n)$, you type:

> **add(*expression of n* , *n = n₀.. n₁*);**

For example,

> **add(n^2, n=1..20);**   # computes $\sum_{n=1}^{20} n^2$.
>
> $$2870$$
>
> **add(sin(n)/n, n=1..5);**
>
> $$\sin(1) + \frac{\sin(2)}{2} + \frac{\sin(3)}{3} + \frac{\sin(4)}{4} + \frac{\sin(5)}{5}$$
>
> **evalf(%);**
>
> $$.9621742221$$

The sum Command

The **add** command works only for finite sums of terms that have numeric values. If the summation involves a parameter (either the number of terms is not explicitly stated or the terms depend on a parameter) this approach is of little help. The **sum** command attempts to replace the summation with a symbolic expression. For example, $\sum_{n=0}^{k} r^n = \frac{r^{k+1}-1}{r-1}$ is a partial sum for a geometric series:

> **sum(r^n, n=0..k);**   # use **sum** for symbolic sums
>
> $$\frac{r^{k+1}}{r-1} - \frac{1}{r-1}$$
>
> **add(r^n, n=0..k);**   # **add** may fail for symbolic sums
>
> `Error, unable to execute add`

Infinite Series

The **sum** command can also be used to find the value of certain infinite series. For example, $\sum_{n=1}^{\infty} \frac{1}{n^2}$ is known to converge:

```
sum( 1/n^2, n=1..infinity );
```

$$\frac{1}{6}\pi^2$$

On the other hand, the harmonic series $\sum_{n=1}^{\infty} \frac{1}{n}$ diverges:

```
sum( 1/n, n=1..infinity );
```

$$\infty$$

Some Famous Infinite Series

Example. Here are three well-known infinite series:

$$\sum_{n=0}^{\infty} \frac{1}{n!} = e, \qquad \sum_{n=0}^{\infty} \frac{1}{(2n+1)^2} = \frac{\pi^2}{8}, \quad \text{and} \quad \sum_{k=1}^{\infty} \frac{k}{e^{2\pi k}-1} = \frac{1}{24} - \frac{1}{8\pi}.$$

Maple recognizes the first two symbolically but can only compute the third numerically.

```
sum( 1/n!, n=0..infinity );
```

$$e$$

```
sum( 1/(2*n+1)^2, n=0..infinity );
```

$$\frac{1}{8}\pi^2$$

```
evalf( sum( k/(exp(2*Pi*k) -1), k=1..infinity ));
```

$$.001877930894$$

```
identify( % );    # find closed form for this decimal
```

$$\frac{1}{24} - \frac{1}{8\pi}$$

Taylor Series

Taylor Polynomials

Recall that if a function f satisfies certain reasonable conditions, then it can be approximated near a point $x = a$ by a polynomial $p_n(x)$ of degree n defined by:

$$p_n(x) = f(a) + \frac{f'(a)}{1!}(x-a) + \frac{f''(a)}{2!}(x-a)^2 + \cdots + \frac{f^{(n)}(a)}{n!}(x-a)^n$$

The polynomial $p_n(x)$ is called the **Taylor polynomial of f of degree n about** $x = a$.

We can use the **add** command to write Taylor polynomials explicitly. For example, e^{2x} has the following sixth-degree Taylor polynomial about the origin:

```
f := x -> exp(2*x);
f(0)+add( eval(diff(f(t),t$k),t=0) *x^k/k!, k=1..6 );
```

$$1 + 2x + 2x^2 + \frac{4}{3}x^3 + \frac{2}{3}x^4 + \frac{4}{15}x^5 + \frac{4}{45}x^6$$

The taylor and convert Commands

Instead of using the clumsy construction shown above, the **taylor** and **convert** commands can produce the Taylor polynomial of degree $n-1$ about $x = a$:

> **convert(taylor(** *function***, x = a, n), polynom);**

For example,

convert(taylor(cos(x), x = 0, 3), polynom);

$$1 - \frac{1}{2}x^2$$

> **Note**: We used the number **3** in the above command to give us the *second* degree polynomial.

convert(taylor(cos(x), x = 0, 8), polynom);
 #Taylor polynomial of degree 7

$$1 - \frac{1}{2}x^2 + \frac{1}{24}x^4 - \frac{1}{720}x^6$$

convert(taylor(cos(x), x=1/2, 3), polynom);

$$\cos\left(\frac{1}{2}\right) - \sin\left(\frac{1}{2}\right)\left(x - \frac{1}{2}\right) - \frac{1}{2}\cos\left(\frac{1}{2}\right)\left(x - \frac{1}{2}\right)^2$$

The **taylor** command by itself actually gives a Taylor polynomial together with a remainder term of degree n, $O\big((x-a)^n\big)$.

taylor(cos(x), x = 0, 8);

$$1 - \frac{1}{2}x^2 + \frac{1}{24}x^4 - \frac{1}{720}x^6 + O\big(x^8\big)$$

Applying **convert** with the **polynom** option to this result drops off the remainder term and returns the Taylor polynomial of degree $n-1$.

convert(%, polynom);

$$1 - \frac{1}{2}x^2 + \frac{1}{24}x^4 - \frac{1}{720}x^6$$

More Examples

Comparing Function and Taylor Polynomials

Example. Consider the function $f(x) = e^{-0.3x} + \sin(x)$:

```
f := x -> exp(-0.3*x) + sin(x):
pict1 :=  plot(f(x), x=-3..3, thickness = 3,
          legend="y=f(x)" ):
```

Suppose we compare *f* graphically with some of its Taylor polynomials near, say, $x = 1$. We start with the Taylor polynomial of degree 3 (but we will not show you the picture right away).

```
p3 := x -> convert( taylor(f(x), x=1, 4), polynom ):
pict3:= plot( p3(x), x=-3..3, legend="y=p3(x)" ):

with(plots):
display([pict1,pict3], legendstyle=[location=right]);
```

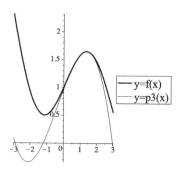

Next we increase the degree of the Taylor polynomial to 5 and then 9:

```
p5 := x -> convert(taylor(f(x), x=1, 6),polynom):
pict5 := plot( p5(x), x=-3..3, color = blue,
        legend="y=p5(x)" ):

p9 := x -> convert(taylor(f(x), x=1, 10),polynom):
pict9 := plot( p9(x), x=-3..3, color = green,
        legend="y=p9(x)" ):

display([pict1, pict5, pict9],
        legendstyle=[location=right]);
```

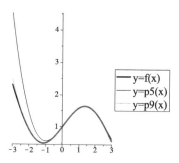

Clearly, the approximations improve as we increase the degree of the polynomial.

Interval of Convergence

Example. Consider the function $g(x) = \dfrac{1}{1+x^2}$:

```
g := x -> 1/(1+x^2):
```

The theory of infinite series tells us that the Taylor series for $g(x)$ near $x = 1$ only converges on a certain interval centered at 1. We would like to see this result with Maple.

We first set up graphs of the function *g* and of its Taylor polynomial approximations.

```
gPict := plot(g(x), x=-2..4, y=-2..3, thickness=3):

g3 := x -> convert(taylor(g(x), x=1, 4),polynom):
gPict3 := plot(g3(x), x=-2..4, y=-2..3, color=blue):

g9 := x -> convert(taylor(g(x), x=1, 10),polynom):
gPict9:= plot(g9(x), x=-2..4, y=-2..3, color=red):

g19 := x -> convert(taylor(g(x), x=1, 20),polynom):
gPict19:= plot(g19(x), x=-2..4,y=-2..3, color=green):
```

Now we use the **display** command to see the graphs together.

```
display( [gPict, gPict3, gPict9, gPict19] );
```

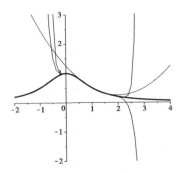

Notice that increasing the degree of the Taylor polynomial improves the approximation on a certain interval around $x = 1$. However, the polynomials turn away sharply at points outside this interval of convergence, say at $x = 2.2$. This cannot be improved no matter the degree of the Taylor polynomials.

```
g99 := x -> convert(taylor(g(x), x=1, 100), polynom):
gPict99:=plot( g99(x), x=-2..4, y=-2..3, color=blue):

display( [gPict, gPict99] );
```

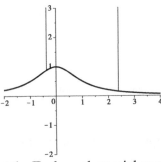

We can tell from the picture that the Taylor polynomial appears to converge to *g* only in the interval $-0.4 \le x \le 2.4$. One can compute theoretically that the interval of convergence is $1 - \sqrt{2} \le x \le 1 + \sqrt{2}$.

Troubleshooting Q & A

Question... Why, when summing a series, do I get strange answers that involve "Psi" or "Ψ"?

Answer... These are special functions that appear quite often in the summation of a series. Their values are well known. For example, you might see:

sum(1/n^2, n=1..1000);

$$-\Psi(1,\ 1001)+\frac{1}{6}\pi^2$$

evalf(%);

1.643934568

Question... When I calculated the numerical value of an infinite series using **evalf** and **sum**, Maple returned the series without evaluation. Does this mean that the series diverges?

Answer... No. This only means that Maple failed to give you an answer. For example,

$\displaystyle\sum_{n=1}^{\infty}\frac{\cos(n)}{n^{3/2}}$ converges while $\displaystyle\sum_{n=1}^{\infty}\cos(n)$ diverges, but both will get a similar response from Maple.

evalf(sum(cos(n)/n^(3/2), n=1..infinity));

$$\sum_{n=1}^{\infty}\frac{\cos(n)}{n^{3/2}}$$

evalf(sum(cos(n), n=1..infinity));

$$\sum_{n=1}^{\infty}\cos(n)$$

Question... How do I decide when to use **add** and when to use **sum**?

Answer... The **sum** command generally tries to find a closed-form expression for the sum. For sums involving an infinite or indefinite number of terms, **add** will not work and **sum** is the only option.

sum(1/(k*(k+1)), k=m..n);

$$-\frac{1}{n+1}+\frac{1}{m}$$

add(1/(k*(k+1)), k=m..n);

```
Error, unable to execute add
```

Question... Which is faster, **sum** or **add**?

Answer... For a small or even moderate number of terms, **add** is almost always faster than **sum**.

```
time( sum( sin(n)/n, n=1..100000 ) );
```

$$4.625$$

```
time( add( sin(n)/n, n=1..100000 ) );
```

$$3.781$$

One exception to this is when there are many terms in the summation *and Maple is able to evaluate the summation in a closed-form*. In such cases, **sum** is much faster.

```
time( sum( 1/(n*(n+1)), n=1..100000 ) );
```

$$0.$$

```
time( add(1/(n*(n+1)), n=1..100000 ) );
```

$$0.520$$

Question... I found a Taylor series using the **taylor** command, but I had trouble plotting it. What should I check for?

Answer... The **taylor** command gives both a Taylor polynomial and a formal error term $O((x-a)^n)$. The **plot** command does not work because of this term. You have to use **convert** to drop this term before you can **plot** it. For example:

```
plot( taylor(sin(x), x=0, 6), x=-3..3);
```

```
Warning, unable to evaluate the function to numeric ....
```

Instead, you should use (plot omitted):

```
plot( convert(taylor(sin(x), x=0, 6), polynom),
        x=-3..3);
```

Question... When I tried plotting a function and a Taylor polynomial on the same graph, the function disappeared. What happened?

```
plot( [ sin(x),
        convert(taylor(sin(x),x=0, 6), polynom) ],
        x= -10..10);
```

Answer... First check the plots separately. If they both work, look at the scales on the *y*-axis. If the scales are very different, one of the graphs may appear as a straight line over the *x*-axis when you put the two curves together. Look at the scale on the *y*-axis. The large *y* scale will make most interesting features of your original function disappear. Specify a range for the *y*-axis:

```
plot( [ sin(x),
        convert(taylor(sin(x),x=0, 6), polynom) ],
        x= -10..10, y= -2..2);
```

CHAPTER 15

Analyzing Differential Equations

Symbolic Solution of ODE

The dsolve Command

The **dsolve** command can be used to find exact (symbolic) or approximate (numeric) solutions to ordinary differential equations. This command is used in the form:

> **dsolve(** *the differential equation* **,** *the unknown function* **);**

> **Note:** The differential equation includes an equal sign **=** and the unknown function is of the form **y(x)**.

Here are some typical examples.

Equation	To Solve It in Maple	Comment
$y' = 3x^2 y$	`dsolve( diff(y(x),x) = 3*x^2*y(x),y(x));` $y(x) = \_C1\, e^{(x^3)}$	`diff(y(x), x)` is used to denote y'.
$y\,y' = -x$	`dsolve(y(x)*diff(y(x),x)=-x, y(x));` $y(x) = \sqrt{-x^2 + \_C1}$, $y(x) = -\sqrt{-x^2 + \_C1}$	There are two solutions for this equation.

$y'' + y' - y = 0$	`dsolve( diff(y(x), x$2) +`
	`        diff(y(x),x) - y(x) = 0,`
	`        y(x));`
	
	$y(x) = \_C1\, e^{\left(-\frac{1}{2}(\sqrt{5}+1)x\right)} + \_C2\, e^{\left(\frac{1}{2}(\sqrt{5}-1)x\right)}$
		This is a *second*-order equation. We use `diff(y(x), x$2)` to denote y''.

$y' = -\cos(xy)$	`dsolve(diff(y(x),x) = -cos(x*y(x)),`
	`        y(x));`
	
	(no output from Maple)
		Maple will not return an output if it cannot solve the differential equation.

> **Note:** $\_C1$ and $\_C2$ denote arbitrary constants in the solutions above.

Equations with Initial or Boundary Conditions

Sometimes you need to solve a differential equation subject to initial or boundary conditions. In such cases, the format of the **dsolve** command is:

> **dsolve(** {*diff equation* **,** *initial / boundary condition(s)* }**, y(x));**

The following table gives several examples of the use of the **dsolve** command with initial conditions:

Initial Value Problem	To Solve It in Maple	Comment

$y' = \dfrac{x}{y}$, with $y(0) = 2$	`dsolve( {diff(y(x),x) = x/y(x), y(0)=2},`
	`        y(x));`
	
	$y(x) = \sqrt{x^2 + 4}$
		None

$y'' - 2y' + y = x$ with $y(0) = 0$, $y'(0) = 2$	`dsolve( {diff(y(x),x$2)-2*diff(y(x),x)+y(x)`
	`        = x,`
	`        y(0)=0, D(y)(0)=2 }, y(x) );`
	
	$y(x) = -2e^x + 3xe^x + 2 + x$
		$y'(0) = 2$ is entered as **D(y)(0)=2**.

The next table shows three typical examples for solving second-order boundary value problems (note that boundary value problems might not have any solution, nor might they have a unique solution).

Boundary Value Problem	To Solve It in Maple	Comment

$y'' + y = 0$ with $y(0) = 0$, $y(\dfrac{\pi}{2}) = 2$	`dsolve( { diff(y(x),x$2) + y(x) = 0,`
	`        y(0)=0, y(Pi/2)=2 }, y(x) );`
	
	$y(x) = \sin(x)$
		unique solution

$y'' + y = 0$ with $y(0) = 0$, $y(\pi) = 2$	`dsolve( { diff(y(x),x$2) + y(x) = 0,`
	`        y(0)=0, y(Pi)=2 }, y(x) );`
	(no output from Maple)
		no solution

$y'' + y = 0$ with $y(0) = 0$, $y(\pi) = 0$	`dsolve( { diff(y(x),x$2) + y(x) = 0,`
	`        y(0)=0, y(Pi)=0 }, y(x) );`
	
	$y(x) = \_C1\sin(x)$
		infinite number of solutions

Numeric Solution of ODE

The numeric Option for dsolve

You can find a numerical approximation to a solution of an initial value problem by adding the **numeric** option in the **dsolve** command. You use it in the following format:

> **dsolve(** *{diff equation, initial condition(s)},* *unknown function,*
> **numeric);**

For example, you can find a numerical approximation for the solution to the equation $y' = -xy$, subject to the initial condition $y(0) = 1$, with:

> **soln := dsolve({diff(y(x),x) = -x*y(x), y(0)=1},**
> **y(x), numeric);**
>
> **proc***(rkf45_x)* ... **end proc**

This output looks strange. Don't worry! Maple reports the answer as a procedure that returns a numerical approximation to the solution. This is easiest to understand with a few examples.

You can compute the values $y(0)$, $y(0.25)$, $y(0.5)$ and $y(1)$ with:

```
soln(0);
```

$$[x=0., y(x) = 1.]$$

```
soln(0.25);
```

$$[x = 0.25, y(x) = 0.969232540034405398],$$

```
soln(0.5);
```

$$[x = 0.25, y(x) = 0.969232540034405398],$$

```
soln(1);
```

$$[x = 1., y(x) = 0.606530524194942355]$$

Or you can see a graph of the solution with the **odeplot** command that's defined in the **plots** package:

```
with( plots ):
odeplot( soln, [x, y(x)], 0..4 );
```

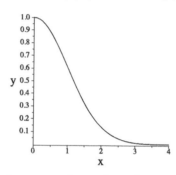

In general, the **odeplot** command can show the graph of the solution $y(x)$ for $a \le x \le b$ with the syntax:

```
with(plots):
odeplot( result from dsolve, [x,y(x)], a..b );
```

Graphical Analysis of ODE

Slope Fields and the DEplot Command

The **DEplot** command can display the slope field for a differential equation $\dfrac{dy}{dt} = f(t, y(t))$. Before using the **DEplot** command, the **DEtools** package must be loaded. The general syntax is:

```
with( DEtools ):

DEplot( diff equation, unknown function, intervals );
```

For example, the slope field for $y' = \dfrac{1}{10}(y - y^2) + \sin(t)$ on the viewing window $0 \le t \le 10$ and $-10 \le y \le 10$ is obtained with:

```
with( DEtools ):
ode := diff(y(t),t)=(y(t)-y(t)^2)/10+sin(t):
DEplot( ode, y(t), t=0..10, y=-10..10 );
```

(We will show you the slope field later.)

You can also add solution curves to the slope field by appending a list of initial conditions $y(t_0) = y_0$ or ordered pair of points $[t_0, y(t_0)]$ that the curves will pass through.

```
DEplot( ode, y(t), t=0..20, y=-10..10,
        [ [y(0)=-2], [0, -1], [0, 2],[5, 5] ] );
```

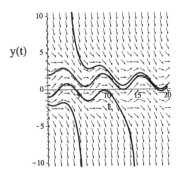

The Useful Tips section of this chapter contains some ideas for ways to improve the appearance of the plots created by **DEplot**.

Systems of ODE

Symbolic Solution of a System of ODE

If you have a system of differential equations involving two (or more) unknown functions, say $x(t)$ and $y(t)$, Maple will attempt to find a symbolic solution when you execute the **dsolve** command in the following format:

> **dsolve[** {*system of diff equations*}, {*unknown functions*} **);**

For example, to solve for functions $x(t)$ and $y(t)$ in the system $x'(t) = x(t) - y(t)$ and $y'(t) = y(t)$, use:

```
dsolve({diff(x(t),t) = x(t) - y(t),
        diff(y(t),t) = y(t)}, {x(t), y(t)});
```

$$\{x(t) = (-\_C2\, t + \_C1)\, \mathbf{e}^t,\, y(t) = \_C2\, \mathbf{e}^t \}$$

Exact and Approximate Solutions to a System of IVPs

The **dsolve** command can be used to solve an initial value problem for a system of differential equations. To solve the system symbolically, the format is:

> **dsolve[** {*system of diff equations, initial condition*},
>   {*unknown functions*} **);**

To solve the system numerically, you will need the **numeric** option.

> **dsolve[** {*system of diff equations*, *initial condition*}**,**
> {*unknown functions*}**, numeric);**

For example, the exact solution to $x'(t) = x(t) - y(t)$ and $y'(t) = y(t)$, with initial conditions $x(0) = 0.2$ and $y(0) = 0.1$, is:

```
dsolve({ diff(x(t),t) = x(t) - y(t),
         diff(y(t),t) = y(t),
         x(0)=0.2, y(0)=0.1 },
       { x(t), y(t) } );
```

$$\left\{ x(t) = \left(-\frac{1}{5}t + \frac{1}{10} \right)\mathbf{e}^t, \ \ y(t) = \frac{1}{5}\mathbf{e}^t \right\}$$

For the initial value problem with $x'(t) = -2x(t)^2 - y(t)$ and $y'(t) = x(t) - y(t)$, with initial conditions $x(0)=0.6$ and $y(0)=0.4$, you can obtain and plot its numerical solution with:

```
soln := dsolve( {diff(x(t),t) = - y(t)^2,
                 diff(y(t),t) = x(t)*y(t),
                 x(0)=0.6, y(0)=0.4},
                {x(t), y(t)}, numeric );
```

$\qquad\qquad$ **proc**(*x_rkf45*) ... **end proc**

```
with( plots ):
odeplot( soln, [x(t),y(t)], 0..10,
         scaling=constrained );
```

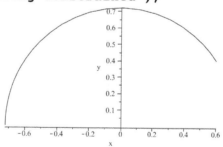

To see the graphs of both $x(t)$ and $y(t)$ plotted against t, you use:

```
odeplot( soln, [[t,x(t)],[t,y(t)]], 0..10,
         legend=["x", "y"] );
```

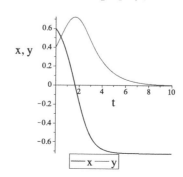

Using DEplot for Systems

The **DEplot** command can be used almost exactly as before to produce a plot of the direction field for a system of ordinary differential equations and/or solution curves for specified initial value problems.

```
with(DEtools):
DEplot( {diff(x(t),t) = - y(t)^2,
         diff(y(t),t) = x(t)*y(t)},
        {x(t), y(t)},
        t=0..5, x=-2..2, y=-2..2,
        [[0,0.9,0.2],[0.,0.95,-1] ] );
```

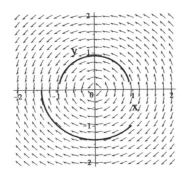

More Examples

More Uses of odeplot

■ **Example.** After using **dsolve** to solve a system of differential equations numerically, you can use **odeplot** to display the plot of any expression involving the solutions $x(t)$ and $y(t)$ against t. For example, to check that the total energy, $\sqrt{x^2(t) + y^2(t)}$, to a system of ordinary differential equations is constant, you can use:

> **soln := dsolve[{***system of diff equations***,*** initial condition***},**
>   **{x(t),y(t)}, numeric):**

> **odeplot(soln, [t, x(t)^2+y(t)^2], 0..10);**
>   (plot omitted)

Equilibrium Analysis for a System

■ **Example.** A standard nonlinear model for a pendulum is:

```
sys := { diff( x(t), t ) = y(t),
         diff( y(t), t ) = -sin(x(t)) };
```

$$\left\{ \frac{d}{dt} x(t) = y(t), \ \frac{d}{dt} y(t) = -\sin(x(t)) \right\}$$

Some of the equilibrium points are at:

> **equilPts := [seq([n*Pi, 0], n = -3 .. 3)];**

$$[[-3\pi,0], [-2\pi,0], [-\pi,0], [0,0], [\pi,0], [2\pi,0], [3\pi,0]]$$

We can draw the direction field, some of the equilibrium solutions, and a few solution curves using the following commands:

```
with(DEtools):
with(plots):

eqPlot:=pointplot(equilPts, symbol = circle,
          symbolsize = 18, color = blue):

ic := [seq([x(0) = 10, y(0) = n],  n=-4..-1),
       seq([x(0) = -10, y(0) = n], n= 1.. 4),
       seq([x(0) = 2*n, y(0) = 0], n=-4.. 4)]:

phasePort := DEplot(sys, {x(t), y(t)},
 t = 0 .. 15, x = -10 .. 10, y = -5 .. 5,
 ic, linecolor = cyan, scaling = constrained):

display(eqPlot, phasePort);
```

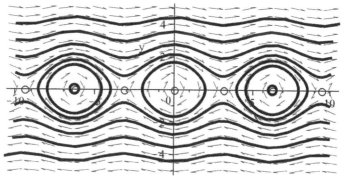

Useful Tips

💡 💡 When using the **dsolve** command with the **implicit** option, Maple will return a symbolic solution in implicit form. For example:

```
dsolve( y(x)*diff(y(x),x) = -x, y(x), implicit );
```

$$y(x)^2 + x^2 - \_C1 = 0$$

💡 💡 The **DEtools** package has several other commands in addition to **DEplot** that can be useful in analyzing an ODE. You can use the examples in the online help to explore them.

💡 💡 For solving partial differential equations, check the on-line help pages for the **PDEtools** package.

You can use the option **dirfield** $=[a, b]$ to control the number of arrows (or lines) for the slope or direction field in **DEplot**. The default value is **[25,25]**. Likewise, you can use the option **stepsize** $= c$ to specify the stepsize used by **DEplot** in the numerical approximation of the solution to an initial value problem.

Troubleshooting Q & A

Question... I got an error message from **dsolve**. What should I look for?

Answer... This most likely means that you did not enter the differential equation(s) correctly. The three most common mistakes made are these:

- You did not match (*parentheses*) correctly.
- You did not follow the rules: y has to be typed as **y(x)**, y' is typed as **diff(y(x), x)**, y'' is typed as **diff(y(x), x$2)**, $y(a)=b$ is entered as **y(a)=b**, and $y'(a) = b$ is typed as **D(y)(a) = b**, and so on.
- You assigned values earlier to either the independent variable (**x**) or the function name (**y**). Try **unassign('x', 'y');** and re-execute.

Question... When I used the **numeric** option for **dsolve,** I got an error message about "... initial conditions." What does this mean, and what should I do?

Answer... Make sure you gave the right number of initial conditions in your input. A first-order differential equation needs one initial condition, a second-order differential equation needs two initial conditions, and so on.

CHAPTER 16
Making Graphs in Space

Graphing Functions of Two Variables

The plot3d Command

Plotting the graphs of functions of two variables with Maple parallels the graphing of functions of one variable. The main change is using **plot3d** instead of **plot**.

In the **plot3d** command, you input an expression in terms of the independent variables x and y, and specify bounds for the x- and y-variables as $x_0 \leq x \leq x_1$ and $y_0 \leq y \leq y_1$. The **plot3d** command then has the form:

> **plot3d(** *an expression of x and y*, **x** = x_0.. x_1, **y** = y_0.. y_1 **);**

For example, the surface whose height is $z = 4 - x^2 - y^2$ above the xy-plane, over the rectangle $-2 \leq x \leq 2$ and $-2 \leq y \leq 2$, is seen with:

```
plot3d( 4-x^2-y^2, x = -2..2, y = -2..2);
```

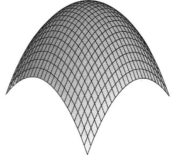

■ **Example.** The graph of the function $f(x,y) = x^2 - y^2$ looks like a saddle.

```
plot3d( x^2 - y^2, x = -3..3 , y = -3..3 );
```

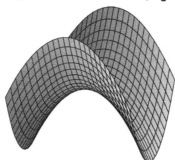

Changing the 3-D View and options from the tool bar and menu bar

Move the cursor to any point in Maple's graphical output, then click and hold the (left) mouse button. As you drag the mouse with the button held down, the picture rotates. This allows you to see the picture from any viewpoint.

When you click on the picture the context bar changes, showing the viewing angles and a series of drop down menus for viewing options. You can use these controls to choose the type of display you want.

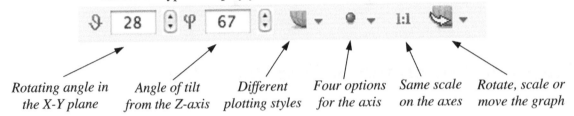

Rotating angle in the X-Y plane   *Angle of tilt from the Z-axis*   *Different plotting styles*   *Four options for the axis*   *Same scale on the axes*   *Rotate, scale or move the graph*

Clicking on a 3-D graph also changes the menu bar at the top of the page, activating the settings in the **Plot** menu. The menus allow you to change the settings for several options, including **Style**, **Symbol**, **Line**, **Color**, **Axes**, **Legend**, **Projection**, and **Lighting**

For example, if you create a picture using **plot3d**, you may want to use the **Ranges** option under the **Axes** menu to adjust the intervals for the *x*, *y* and *z* axes. (See the picture on the right.)

Specifying Options

The grid Option

The **grid** option is used to instruct **plot3d** to sample more points when constructing a plot. Consider:

```
plot3d( sin(3*y)+cos(5*x), x = -Pi ..Pi, y=-Pi..Pi );
```

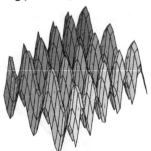

The picture does not look very good. But if we add the option **grid = [50,50]**, we'll see the graph shown with values sampled from a 50×50 grid, instead of the default 25×25 grid. This gives a smoother picture, but takes about four times longer to compute and draw.

```
plot3d( sin(3*y)+cos(5*x), x = -Pi ..Pi, y=-Pi..Pi,
        grid=[50,50] );
```

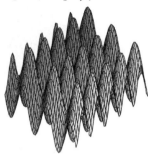

The view Option

As with graphs of functions of one variable, we sometimes need to control the range of the dependent variable in 3-D graphs. With the **plot3d** command we use the **view** option. For example, if we try the graph

```
plot3d(1/(x^2+y^2), x=-2..2, y=-2..2);
```

The result is not very informative because the values of this function close to the origin are very large and distort the scale of the z-axis. If you try,

```
plot3d(1/(x^2+y^2), x=-2..2, y=-2..2, z=-5..5);   #error
```

you will get an error message. However, you can use the **view** option as follow

```
plot3d(1/(x^2+y^2), x=-2..2, y=-2..2, view =-5..5);
```

This truncated graph enables you to see more of the detail in the surface.

Other Useful Options

The following table summarizes some of the more common options that you can use to add more "character" to a picture with **plot3d**.

Option	*What It Does*
`axes = normal`	Shows the three axes in a normal way.
`axes = boxed`	Surrounds the picture with a box.
`labels = ["x","y","z"]`	Provides names to label each of the three axes.
`scaling = constrained`	Scales the graphic so that units in each direction have the same length.
`color = `*the color you want*	Draw the graphic with the specified color.

■ **Example.** The sombrero has equation $f(x,y) = \sin(\sqrt{x^2 + y^2})/\sqrt{x^2 + y^2}$:

```
S := sin(sqrt(x^2+y^2))/sqrt(x^2+y^2):
plot3d( S, x=-7..7, y=-7..7, grid=[50,50], color=red,
    axes=boxed, labels=["x axis","y axis","z axis"]);
```

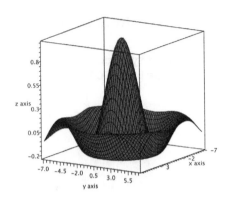

Surfaces in Cylindrical and Spherical Coordinates

Surfaces are sometimes described using the cylindrical or spherical coordinate system. Maple can draw such surfaces easily by setting the **coords** option to **cylindrical** or **spherical**, respectively.

The coords = cylindrical Option

Points in the cylindrical coordinate system are described by quantities r, θ, and z, where

- r is the horizontal radial distance of the point from the z-axis;
- θ is the horizontal angle measured from the positive x-axis; and
- z is the vertical distance from the xy-plane.

To draw the surface $r = f(\theta, z)$ for $\theta_0 \le \theta \le \theta_1$ and $z_0 \le z \le z_1$, you enter:

> ```
> plot3d( f(θ, z), theta = θ₀..θ₁, z = z₀..z₁
>         coords = cylindrical);
> ```

> **Note:** When using **coords=cylindrical**, you must enter the interval for **theta** first, then the interval for **z**. You do not have to use these names, but Maple assumes this ordering when it creates the plot.

For example, to see the surface $r = z^2 \cos^2(3\theta)$, for $0 \le \theta \le 2\pi$ and $-2 \le z \le 2$:

```
plot3d( z^2*(cos(3*theta))^2, theta=0..2*Pi,
        z=-2..2, coords = cylindrical );
```

Notice that this surface is very "choppy" as the horizontal angle θ varies, yet the surface is quite smooth along the vertical z-direction. This suggests that we should increase the sampling of points in the θ variable (from 25 to, say, 80) but decrease the sampling in z (from 25 to, say, 15). We can do this with:

```
plot3d( z^2*(cos(3*theta))^2, theta=0..2*Pi,
        z=-2..2, coords=cylindrical, grid=[80,15]);
```

The coords = spherical Option

Points in the spherical coordinate system are described by quantities ρ, θ, and ϕ, where

- ρ is the radial distance in space of the point from the origin;
- θ is the horizontal angle measured from the positive x-axis; and
- ϕ is the vertical angle measured from the positive z-axis.

To draw the surface $\rho = f(\theta,\phi)$, $\theta_0 \leq \theta \leq \theta_1$, and $\phi_0 \leq \phi \leq \phi_1$, you will enter:

```
plot3d( f(θ, φ), theta = θ₀..θ₁, phi = φ₀..φ₁,
        coords = spherical );
```

> **Note:** When using **coords=spherical**, you have to enter the interval for **theta** first, then the interval for **phi**.

For example, to see the surface $\rho = \sqrt{\theta}(3 + \cos\phi)$, $0 \leq \theta \leq 3\pi/2$, and $0 \leq \phi \leq \pi$:

```
plot3d( sqrt(theta)*(3+cos(phi)), theta =0..3*Pi/2,
        phi=0..Pi, coords=spherical);
```

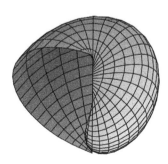

More Examples

Choosing the Right Coordinate Systems

In some cases, a surface given in rectangular coordinates will look better if you draw it using cylindrical or spherical coordinates.

■ **Example.** The surface $z = \dfrac{x^2 - y^2}{(x^2 + y^2)^2}$ can be plotted with:

```
plot3d( (x^2-y^2)/(x^2+y^2)^2, x=-3..3, y=-3..3 );
```

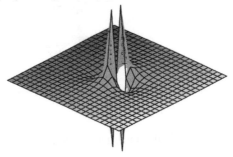

The picture is choppy especially near the origin. However, if we use cylindrical coordinates, the equation of the surface becomes

$$z = \frac{x^2 - y^2}{(x^2 + y^2)^2} = \frac{(r\cos\theta)^2 - (r\sin\theta)^2}{((r\cos\theta)^2 + (r\sin\theta)^2)^2} = \frac{r^2(\cos^2\theta - \sin^2\theta)}{r^4} = \frac{\cos 2\theta}{r^2},$$

which means $r^2 = \cos(2\theta)/z$, or equivalently $r = \sqrt{\cos(2\theta)/z}$. (The surface needs to be expressed in the form $r = f(\theta, z)$ in order to use **coords=cylindrical**.)

```
S := (theta,z) -> sqrt(cos(2*theta)/z):
top := plot3d( S(theta,z), theta=0..2*Pi,
        z=0.1..2, grid=[50,15], coords=cylindrical):
bottom := plot3d( S(theta,z), theta=0..2*Pi,
        z=-2..-0.1, grid=[50,15], coords=cylindrical):
plots[display]( [top,bottom], orientation=[36,77]);
```

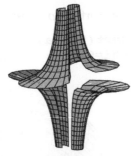

This picture is nicer! Since this surface is most easily described with z as a function of r and θ, the graph is even nicer if we plot it as a parametric surface in cylindrical coordinates. We will do that in chapter 20.

Useful Tips

The **plot3d** command can be used to plot with non-rectangular domains. For example, to graph the part of the surface $z = x^2 + y$ above the region in the xy-plane bounded by $y = x^2$ and $y = 2 - x^2$ use:

```
plot3d( x^2+y, y=x^2..2-x^2, x=-1..1, axes=normal );
```

This approach can be used in other coordinate systems as well. (When viewed from above, **orientation = [0, 0]**, you see the projection of this surface onto the xy-plane.)

For 2-D graphs having multiple curves, Maple automatically uses a different color for each curve. For 3-D graphs, color variations represent depth. This means that if multiple surfaces are graphed together, they will have the same coloring scheme.

To designate separate colors for different surfaces in the same picture, you can either plot each one separately using a different **color** option and then use the **display** command to put them together, or you can use square brackets **[ ]** to make list of expressions to plot and make a corresponding list of colors. For example:

```
pict1 := plot3d(10-x^2, x=-3..3, y=-3..3, color=red):
pict2 := plot3d(y^2, x=-3..3, y=-3..3, color=green):
plots[display]( [pict1, pict2] );
```

or:

```
plot3d( [10-x^2, y^2], x=-3..3, y=-3..3,
        color=[red, green]);
```

If you want to create a 3-D graph of an expression that Maple has produced as output from a previous command, simply position the cursor over this expression and click the output with the right mouse button (or $\boxed{control -}$ click for Mac users). A **context-sensitive menu** pops up. You can then select **Plots → 3D-Plot → x,y**. Maple will automatically show you the picture. If you want to add more surfaces to this plot, simply select the expressions (expressed in x and y) and drag the input into the previous picture.

The interactive **Plot Builder** is useful for learning the syntax of new plotting options. You launch **Plot Builder** by selecting **Tools → Assistants → Plot Builder** from the menu bar. Enter the expressions to be plotted and click OK. Select the type of plot to be created, then click **Options**. Specify the settings that you want, clicking **Preview** to check the current status of the plot before clicking **Plot** to see the final plot. (Clicking **Command** returns the actual Maple plot command used to create the plot without actually showing the plot.)

You can use **plot3d** to draw parametric surfaces in space. We will discuss this in detail in Chapter 20.

Troubleshooting Q & A

Question... I tried to draw a 3-D picture but got an error message telling me about a "Plotting error." What happened?

Answer... This means that Maple has trouble using the function you specified to generate points for the picture. Check these items:

- Did you mistype the input?
- Did you use the same variables in the function as you used in specifying the intervals?
- Is the function well-defined in the given intervals?

Another common mistake is to use the **plot** command to draw a 3-D picture; make sure you use the **plot3d** command.

Question... I drew the graph of a function using **plot3d**, but the picture looks different from the one shown in my textbook. Why is that?

Answer... Two plots may sometimes look different because they are drawn in different viewing windows, with different scalings, or from different perspectives. Adjust the viewing window with the **view**, the **scaling**, and **orientation** options respectively.

Question... Can I use **display** to combine the graphics created using context-sensitive menus?

Answer... You cannot use **display** to combine pictures constructed with context-sensitive menus, including drag-and-drop.

The pictures created from context-sensitive menus are based on the **smartplot3d** command. Although the outputs from **smartplot3d** and **plot3d** appear to be similar, their internal structures are quite different. The reason is that the pictures from **smartplot3d** allow several manipulations that are not offered in the standard plots.

For example, you can control-drag an expression into a **smartplot3d** picture to draw that expression. As a result, we cannot combine these pictures using **display** as we can with standard plots.

Question... The picture I got using **coords = cylindrical** or **coords = spherical** was completely wrong. What should I check?

Answer... Check these three areas:

- Make sure you typed the input function and the intervals of the two parameters correctly.
- With **coords = cylindrical** you have to enter the θ-interval **theta** = $\theta_0 .. \theta_1$ first, followed by the z-interval **z** = $z_0 .. z_1$. If you enter these in the wrong order, Maple will reverse the sense of the variables.
- With **coords = spherical** you have to enter the θ-interval **theta** = $\theta_0 .. \theta_1$ first, then the ϕ-interval **phi** = $\phi_0 .. \phi_1$. If you enter these in the wrong order, Maple will draw an incorrect picture.

CHAPTER 17

Level Curves and Level Surfaces

Level Curves in the Plane

The contourplot and contourplot3d Commands

In Maple, the level curves (contours) of a function $f(x, y)$ are plotted with the **contourplot** command that is defined in the **plots** package. To see the level curves inside the rectangle $x_0 \leq x \leq x_1$, $y_0 \leq y \leq y_1$, you enter:

```
with(plots);
contourplot( function , x= x0..x1, y= y0..y1);
```

(This syntax looks just like the **plot3d** command syntax that we discussed in the previous chapter.) For example, here are some level curves of $f(x, y) = x y e^{-x^2 - y^2}$:

```
contourplot(x*y*exp(-x^2-y^2), x=-2..2, y=-2..2);
```

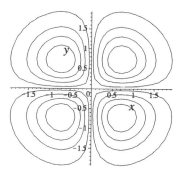

The **contourplot3d** command accepts the same arguments and options as **contourplot**. The differences are that **contourplot3d** produces a 3-D view of the contours raised to their appropriate levels and **contourplot3d** is faster.

```
contourplot3d(x*y*exp(-x^2-y^2), x=-2..2, y=-2..2);
```

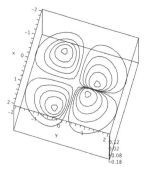

Options for Contour Plotting

Some of the options we like to use with **contourplot** and **cotourplot3d** include the following:

Option	What It Does
contours = *n*	Draws *n* level curves
grid = [*n*, *m*]	Changes resolution of the picture
filled = true	Shades the areas between level curves.
coloring = [white, blue]	Lighter blues represent lower levels, while darker blues represent higher levels.
scaling = constrained	Both the *x*- and *y*-axes are drawn with the same scale.

Here's a nicer picture than the one shown earlier:

```
contourplot( x*y*exp(-x^2-y^2), x=-2..2, y=-2..2,
        contours=20, grid=[35,35], filled=true,
        coloring=[white,blue], scaling=constrained);
```

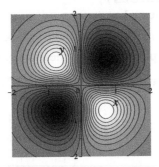

With our choice of **coloring**, lighter shades represent lower levels, while darker shades represent higher levels. We can tell from this picture that the function has its largest values near $(0.8, 0.8)$ and $(-0.8, -0.8)$.

Plotting Specific Level Curves

You can plot specific level curves using the **contours** option in the form **contours = [*the levels*]**. The levels must be separated by commas. For example, to see the contours at levels $0, 0.1,$ and 0.15 without any shading between the contours:

```
contourplot(x*y*exp(-x^2-y^2), x=-2..2, y=-2..2,
        grid = [ 30,30 ], contours = [0,0.1,0.15]);
```

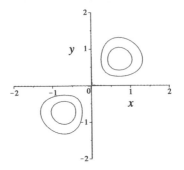

Notice that the level curve of $f(x, y) = x y e^{-x^2 - y^2}$ at level 0 consists of the x- and y-axes. (Maple drew the level curve with squiggles near the origin.)

Level Surfaces in Space

The implicitplot3d Command

If $f(x, y, z)$ is a function of three variables defined over a rectangular region $x_0 \leq x \leq x_1$, $y_0 \leq y \leq y_1$ and $z_0 \leq z \leq z_1$, then the level surface of f at level c can be seen with the **implicitplot3d** command. It's defined in the **plots** package.

```
with(plots);
implicitplot3d( f(x, y, z) = c,
            x = x_0 .. x_1,  y = y_0 .. y_1,  z = z_0 .. z_1 );
```

You can draw more than one level surface in the same graphic, say with levels c_1 and c_2, by entering:

```
implicitplot3d( [f(x, y, z) = c_1 ,  f(x, y, z) = c_2],
            x = x_0 .. x_1,  y = y_0 .. y_1,  z = z_0 .. z_1 );
```

Here are the level surfaces of $f(x, y, z) = x^3 - y^2 + z^2$ at the levels 1 and 10:

```
f := (x,y,z) -> x^3-y^2+z^2 ;
with(plots):
implicitplot3d( [ f(x,y,z)=1, f(x,y,z)=10 ],
            x=-2..5, y=-2..2, z=-2..3);
```

In other words, the surfaces shown above have equations $x^3 - y^2 + z^2 = 1$ and $x^3 - y^2 + z^2 = 10$.

More Examples

Comparing plot3d with contourplot

■ **Example.** Consider $f(x, y) = x^2 - y^2$. The following command will show the contours at level 0, 1, and –1.

```
contourplot( x^2- y^2, x=-2..2, y=-2..2,
            contours=[0, 1, -1], scaling=constrained);
```

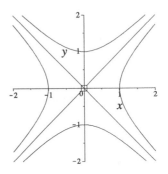

What are these curves? Recall that a contour at level c is given by the equation
$f(x, y) = c$. Thus points on the contours at level 0, 1, and –1 satisfy $x^2 - y^2 = 0$,
$x^2 - y^2 = 1$, and $x^2 - y^2 = -1$, respectively. These are, in order, the two lines
$y = \pm x$, a hyperbola that opens left/right, and a hyperbola that opens up/down.

These contours can be seen in the 3-D plot with **orientation=[-90,0]**. (From
above, with the x- and y- axes in standard position.) For example we see the center
contour by intersecting the graphs of $f(x,y) = x^2 - y^2$ and $z = 0$. To make the
intersection clear we use **color** options in the individual plots we make, and then
combine them with the **display** command.

```
P := plot3d( x^2-y^2, x=-2..2, y=-2..2, color=white):
level0 := plot3d( 0, x=-2..2, y=-2..2, color=grey):
plots[display]( [P, level0], orientation =[-90,0]);
```

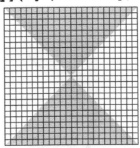

You can also see the relationship between level curves and the graph by using the
style=patchcontour option on the **plot3d** command. Type:

```
plot3d( x^2-y^2, x=-2..2, y=-2..2,
        style=patchcontour, contours=[0, 1, -1],
        axes=normal, orientation=[-90,0] );
```

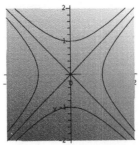

You see the level curves here. Both of these pictures are 3-D objects. Now, click on a plot and rotate the frame box to another position (**orientation=[60,60]**):

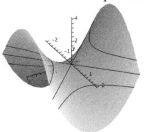

You can now see how the level curves are formed from the graph.

Curves in the Plane Defined by an Equation

■ **Example**. The equation $2x^2 - 3xy + 5y^2 - 6x + 7y = 8$ defines a rotated ellipse in the plane. We could use **implicitplot** to draw it. But it's also just the level curve of the function $f(x, y) = 2x^2 - 3xy + 5y^2 - 6x + 7y$ at level 8. We can see it with **contourplot**:

```
contourplot( 2*x^2-3*x*y+5*y^2-6*x+7*y, x=-2..5,
             y=-3..2, contours=[8], grid=[30,30] );
```

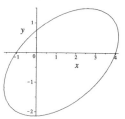

(The intervals used in the command above were arrived at after some trial-and-error, so that we could give you a nice picture.)

The Quadric Surfaces in Space

The quadric surfaces are those surfaces in space that can be given by an equation of the form:

$$Ax^2 + By^2 + Cz^2 + Dxy + Eyz + Fxz + Gx + Hy + Iz + J = 0,$$

where $A, B, C, \ldots, J$ are constants. These surfaces are discussed in detail in many multivariable calculus books. With the help of the **implicitplot3d** command, we can easily see pictures of various quadric surfaces.

■ **Example**. The equation $\dfrac{x^2}{2^2} + \dfrac{y^2}{3^2} - \dfrac{z^2}{4^2} = 1$ defines a hyperboloid of one sheet:

```
implicitplot3d( x^2/2^2 + y^2/3^2 - z^2/4^2 = 1,
                x=-10..10, y=-10..10, z=-10..10 );
```

Troubleshooting Q & A

Question... When I tried **contourplot** or **implicitplot3d**, Maple returned the command to me unevaluated. What should I check?

Answer... Most likely you did not load the **plots** package before using the command. Type:

```
with(plots);
```

Now, try the command again.

Question... I got an empty picture from **contourplot** when I specified a list of levels in the **contours** option. What happened?

Answer... Most likely the level curve(s) lie completely outside the specified viewing window. For example, the contour $x^2 + y^2 = 9$ contains the rectangle $-2 \le x \le 2$, $-2 \le y \le 2$, so you get an empty picture when you enter:

```
contourplot( x^2+y^2, x=-2..2, y=-2..2,
                       contours = [9]);
```

Question... I got an empty plot from **implicitplot3d**. What happened?

Answer... There are four likely possibilities:

- The level surface does not exist, so nothing can be shown in the output (e.g., the level surface $x^2 + y^2 + z^2 = -1$).

- The level surface you're trying to see doesn't lie inside the region $x_0 \le x \le x_1$, $y_0 \le y \le y_1$, and $z_0 \le z \le z_1$ that you gave. (This is similar to the problem addressed in the question above.) Recheck both the level and the region you specified.

- The region you specified may be too large for the level surface to be noticeable. For example:
```
implicitplot3d( x^2+y^2 = 1, x=-10..10,
                y=-10..10, z=-10..10);    #no picture
implicitplot3d( x^2+y^2 = 1, x =-2..2,
                y=-2..2, z=-2..2);        #shows a picture
```

- The level surface is not entered in the form $f(x,y,z) = c$, where c is the level of the level surface.

CHAPTER 18
Partial Differentiation and Multiple Integration

Partial Derivatives

The diff Command

The **diff** command we used in Chapter 12 is actually a partial differentiation operator. It differentiates an expression with respect to a specified variable, treating all other symbols as constants.

> **diff(** *the function or expression* , *variable* **);**

For example, if $f(x, y) = 3xy^2 - 5y\sin x$, its partial derivatives $f_x = \dfrac{\partial f}{\partial x}$ and $f_y = \dfrac{\partial f}{\partial y}$ are computed with:

```
f := (x,y) -> 3*x*y^2 - 5*y*sin(x):
diff( f(x,y), x );
```
$$3y^2 - 5y\cos(x)$$

```
diff( f(x,y), y );
```
$$6xy - 5\sin(x)$$

You find higher-order derivatives by listing the variables in the order of differentiation. For example, $f_{xxy} = \dfrac{\partial^3 f}{\partial y \partial x^2}$ represents taking the partial derivative "first by x, then by x, then by y." You compute this with:

```
diff( f(x,y), x, x, y );
```
$$5\sin(x)$$

Double and Triple Integrals

Iterated Double Integral

There are two ways to compute the iterated double integral $\int_a^b \int_{g_1(x)}^{g_2(x)} f(x, y)\, dy\, dx$.

The easiest way is to use the **MultiInt** command from the **Student[MultivariateCalculus]** package. It is computed with:

```
with( Student[MultivariateCalculus] ):
MultiInt( f(x, y), y = g₁(x).. g₂(x), x = a..b);
```

The second way is to use the **int** command iteratively,

```
int( int( f(x, y), y = g₁(x).. g₂(x) ), x = a..b);
```

Similarly, the iterated double integral $\int_c^d \int_{h_1(y)}^{h_2(y)} f(x, y)\, dx\, dy$ is computed with:

```
with(Student[MultivariateCalculus]):
MultiInt( f(x, y), x = h₁(y).. h₂(y), y = c..d);
```

or

```
int( int( f(x, y), x = h₁(y).. h₂(y) ), y = c..d);
```

■ **Example.** The iterated integral $\int_0^2 \int_0^{2x} (3x^2 + (y-2)^2)\, dy\, dx$ is computed with:

```
with(Student[MultivariateCalculus]):
MultiInt( 3*x^2 + (y-2)^2,
          y=0..2*x, x=0..2);
```

or

```
int( int( 3*x^2 + (y-2)^2, y=0..2*x),
    x=0..2);
```

$$\frac{88}{3}$$

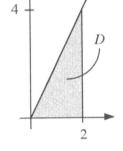

The integration above takes place over the region D defined by the inequalities:

$$0 \le x \le 2 \text{ and } 0 \le y \le 2x.$$

(See the picture to the right.) D can also be described by the inequalities:

$$0 \le y \le 4 \text{ and } y/2 \le x \le 2$$

It follows that the iterated integral $\int_0^2 \int_0^{2x} (3x^2 + (y-2)^2)\, dy\, dx$ has the same value

as $\int_0^4 \int_{y/2}^2 (3x^2 + (y-2)^2)\, dx\, dy$:

```
MultiInt(3*x^2 + (y-2)^2, x = y/2..2, y = 0..4);
```

$$\frac{88}{3}$$

We can see the iterated integral that's being computed by using the **output=integral** option.

```
MultiInt(3*x^2 + (y-2)^2, x = y/2..2, y = 0..4,
         output=integral);
```

or

```
Int( Int( 3*x^2 + (y-2)^2, x=y/2..2x), y=0..4 );
```

$$\int_0^4 \int_{\frac{1}{2}y}^2 \left(3x^2 + (y-2)^2\right) dx\, dy$$

Iterated Triple Integrals

An iterated triple integral $\int_a^b \int_{g_1(x)}^{g_2(x)} \int_{h_1(x,y)}^{h_2(x,y)} f(x, y, z)\, dz\, dy\, dx$ can be evaluated using the **MultiInt** command with three ranges:

```
with(Student[MultivariateCalculus]):
MultiInt( f(x, y, z), z = h₁(x, y).. h₂(x, y),
          y = g₁(x)..g₂(x), x = a..b );
```

or with three nested **int** commands:

```
int( int( int( f(x, y, z), z = h₁(x, y).. h₂(x, y) ),
          y = g₁(x).. g₂(x)), x = a..b );
```

Other variations in the order of integration can be evaluated similarly.

■ **Example**. To evaluate $\int_{-3}^{3} \int_{-\sqrt{9-x^2}}^{\sqrt{9-x^2}} \int_{x+y}^{3+y} z^2\, dz\, dy\, dx$, write:

```
with(Student[MultivariateCalculus]):
MultiInt( z^2, z=x+y..3+y,
         y=-sqrt(9-x^2)..sqrt(9-x^2), x=-3..3);
```

or

```
int( int( int( z^2, z = x+y..3+y),
          y=-sqrt(9-x^2)..sqrt(9-x^2)), x=-3..3 );
```

$$\frac{567}{4}\pi$$

Numerical Integration

If you want to find a numeric approximation for a double or triple integral, you should use **evalf** together with the **MultiInt** command and the **output=integral** option.

■ **Example**. To find $\int_{-3}^{3} \int_{-\sqrt{9-x^2}}^{\sqrt{9-x^2}} \int_{x+y}^{3+y} z^2\, dz\, dy\, dx$ numerically, write:

```
evalf( MultiInt( z^2, z = x+y..3+y,
                 y=-sqrt(9-x^2)..sqrt(9-x^2),
                 x=-3..3, output=integral) );
```

445.3207586

Or you can use **evalf** together with the **Int** command (on all 3 integrals):

```
evalf( Int( Int( Int( z^2, z=x+y..3+y ),
                 y=-sqrt(9-x^2)..sqrt(9-x^2) ),
            x=-3 ..3 ) );
```

445.3207586

As you may expect, numerical integration will give you an answer quickly in most cases, and it can even be used when symbolic integration fails.

More Examples

Critical Points and the Hessian Test

■ **Example.** Suppose that $f(x,y) = x^4 - 3x^2 - 2y^3 + 3y + 0.5xy$. We can find its critical points by solving the equations $f_x = 0$ and $f_y = 0$ simultaneously:

```
f := (x,y) -> x^4 - 3*x^2 - 2*y^3 + 3*y + 0.5*x*y:
critPt := [ solve( { diff(f(x,y),x) = 0,
   diff(f(x,y),y)=0 }, {x,y} ) ];
```

$[\ \{\ x = -1.250162405, y = 0.6291421140\ \}, \{\ x = 0.05935572277, y = 0.7105957432\ \},$
$\{\ x = 1.191117672, y = 0.7741187286\ \}, \{\ x = -1.197482099, y = -.6326213916\ \},$
$\{\ x = -0.05877159444, y = -0.7036351094\ \}, \{\ x = 1.255942703, y = -0.7776000848\ \}\ ]$

There are six critical points. We will define the discriminant

$$D = (f_{xx})(f_{yy}) - (f_{xy})^2$$

and evaluate it at each critical point. The Hessian Test you learned in multivariable calculus says:

- If $D < 0$, then the critical point is a saddle point.
- If $D > 0$ and f_{xx} is negative, then the critical point is a local maximum.
- If $D > 0$ and f_{xx} is positive , then the critical point is a local minimum.

We compute the discriminant D and f_{xx} at each of the critical points:

```
dis := diff( f(x,y), x,x )*diff( f(x,y), y,y )
       - diff( f(x,y), x,y )^2;
```

$$-12\ (12\ x^2 - 6)\ y - 0.25$$

```
eval([dis, diff(f(x,y),x,x)], critPt[1] );
```

$$[-96.54552914,\ 12.75487247]$$

This means that the first critical point is a saddle point. We can check all the critical points at once with:

```
seq(eval([dis,diff(f(x,y),x,x)], critPt[n]), n=1..6);
```

$[-96.54552914,\ 12.75487247],\quad [50.55238934, -5.957722778],$
$[-102.6671685,\ 11.02513571],\quad [84.83171040,\ 11.20756052],$
$[-50.56174649, -5.958550796],\quad [120.3903441,\ 12.92870488]$

This result shows that the discriminant is positive at the second, fourth, and sixth of the critical points. Since f_{xx} is positive at the fourth and sixth critical points, $(-1.197, -0.633)$ and $(1.256, -0.778)$ will be local minima for f. Also, the point $(0.059, 0.711)$ will be the local maximum.

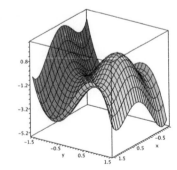

The three remaining critical points have negative discriminants, so each is a saddle point for f. You can check these points from the graph of f shown on the right.

Method of Lagrange Multipliers

■ **Example**. We want to find the maximum and minimum values of the function $f(x, y) = (x-2)y + y^2$, subject to the constraint $x^2 + y^2 - 1 = 0$. The Method of Lagrange Multipliers states that we need to solve the system of equations:

$$f_x = \lambda g_x, \quad f_y = \lambda g_y \text{ and } g = 0,$$

where g is the "constraint" function $g(x, y) = x^2 + y^2 - 1$.

```
f := (x,y) -> (x-2)*y + y^2:
g := (x,y) -> x^2 + y^2 - 1:
soln := [ solve( { diff( f(x,y),x)=p*diff(g(x,y),x),
                   diff( f(x,y),y)=p*diff(g(x,y),y),
                   g(x,y) = 0.0 }, {p,x,y} ) ]:
```

(Note: We use the variable **p** in this computation to stand for the multiplier λ.)

Since **soln** contains both real and complex solutions, we can use the **remove** command (as mentioned at the end of Chapter 6) to select only the real solutions.

```
realSoln := remove( has, soln, I );
```

$[\{p = 2.143478901, x = -0.227167105, y = -0.9738557934\},$
$\{p = -0.6483805940, x = -0.6106661474, y = 0.7918881588\}]$

We can turn this into a list of points (x, y), and function values $f(x,y)$, with:

```
Q := [seq(eval([[x,y], f(x,y)], realSoln[n]),
     n=1..2)];
```

$[[-0.227167105, -0.9738557934], 3.117334694],$
$[[-0.6106661474, 0.7918881588], -1.440268752]$

So f has the minimum value -1.44027 at the point $(-0.6107, 0.7919)$ and the maximum value 3.11733 at the point $(-0.2272, -0.9739)$.

You can also see this result geometrically by drawing the contour plot of f and the constraining circle $g = 0$ together. But first, we need to get the list of extreme points:

```
solnPts := [Q[1][1], Q[2][1]];
```

$[[-0.227167105, -0.9738557934], [-0.6106661474, 0.7918881588]]$

```
with(plots):
P1 := contourplot( f(x,y), x=-1.5..1.5,
           y=-1.5..1.5, contours=20):
P2 := implicitplot( g(x,y) = 0, x=-2..2,
           y=-2..2, thickness = 3 ):
P3 := pointplot(solnPts, symbol=solidcircle,
           symbolsize = 20):
display( [P1,P2,P3], scaling=constrained );
```

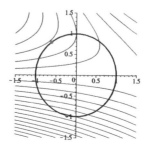

Integration in Polar Coordinates

■ **Example.** Let us compute the integral $\iint_D e^{(x^2+y^2)} dA$, where D is the circular sector sketched to the right. In polar coordinates, D can be described as $0 \leq r \leq 1$, $\pi/4 \leq \theta \leq \pi/2$.

We learned from calculus that we must first write the integral as the iterated integral $\int_{\pi/4}^{\pi/2} \int_0^1 e^{r^2} r \, dr \, d\theta$. Then we can compute its value:

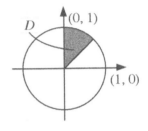

```
MultiInt( exp(r^2)*r, r=0..1,
          t=Pi/4..Pi/2 );
```

$$-\frac{1}{8}\pi + \frac{1}{8}e\pi$$

Outsmarting Maple

■ **Example.** You can outsmart Maple easily. Let D be the region inside the circle $x^2 + y^2 = 1$. Calculate $\iint_D \sqrt[3]{x^2 + y^2} \, dA$ both by hand and using the computer.

• For the computer: Evaluate the double integral using rectangular coordinates. The region D is described by the inequalities $-1 \leq x \leq 1$ and $-\sqrt{1-x^2} \leq y \leq \sqrt{1-x^2}$, so you enter:

```
MultiInt((x^2+y^2)^(1/3),
         y = -sqrt(1-x^2)..sqrt(1-x^2), x=-1..1);
```

After a while, Maple will give you an unevaluated integral.

• However, you can evaluate the integral directly and easily using polar coordinates, because the region D is described by the inequalities $0 \leq r \leq 1$ and $0 \leq \theta \leq 2\pi$:

$$\iint_D \sqrt[3]{x^2 + y^2} \, dA = \int_0^{2\pi} \int_0^1 (r^2)^{1/3} r \, dr \, d\theta = \int_0^{2\pi} \int_0^1 (r^{5/3}) \, dr \, d\theta$$

$$= 2\pi(\frac{3}{8}r^{8/3}\Big|_0^1) = \frac{3\pi}{4}.$$

You Win!!! While Maple is powerful, it does not have any true intelligence.

Is Fubini's Theorem Wrong?

■ **Example.** You may recall that for an iterated integral, the order of integration should not affect the value of the double integral. This is called Fubini's Theorem, which basically says that

$$\int_a^b \int_c^d f(x,y) \, dy \, dx = \int_c^d \int_a^b f(x,y) \, dx \, dy,$$

when f is continuous on the rectangle $a \leq x \leq b$, $c \leq y \leq d$.

However, if you try $f(x,y) = \dfrac{x+y}{x-y}$ on $0 \leq x \leq 1$, $-1 \leq y \leq 0$:

```
MultiInt( (x+y)/(x-y)^3, y=-1..0, x=0..1);
```

$$\frac{1}{2}$$

```
MultiInt( (x+y)/(x-y)^3, x=0..1, y=-1..0);
```

$$-\frac{1}{2}$$

The answers are different, but should you be concerned? No, because the function $\frac{x+y}{(x-y)^3}$ is not continuous on the region of integration (it fails to be continuous on the line $y = x$) and Fubini's Theorem does not apply.

Useful Tips

💡 💡 The **MultiInt** command has an **output=steps** option that gives intermediate results.

```
MultiInt(3*x^2+(y-2)^2, y = 0 .. 2*x,
    x = 0 .. 2, output = steps);
```

Maple shows the initial iterated integral and the result after each antidifferentiation and substitution step.

💡 💡 The numerical evaluation of iterated integrals using **evalf** with **Int** can be slow because Maple is actually computing iterated numerical one-dimensional integrals. The combination of **evalf** with **MultiInt** and **option=integral** is much faster because it uses a true multi-dimensional numerical integration. You can use a combination of **int** and **Int** in an iterated integral, but remember that **int** tries to evaluate the integral symbolically before starting to use a numerical method.

💡 💡 When you use the **solve** command in finding the critical points or the Lagrange multiplier method, Maple may give you the answers in the form "**RootOf**". You can avoid that by adding a decimal point to any one of the integer coefficients of the equations. (In our method of Lagrange Multiplier example earlier, we replaced "0" with "0.0" in the equation.) Maple will then give you the answers in numerical form.

Troubleshooting Q & A

Question... I entered a double (or triple) integral expression and have been waiting for a response for two or three minutes. Maple hasn't given me an answer yet. Is something wrong?

Answer... Integration is a difficult mathematical problem. Unfortunately, Maple cannot solve every integration problem you encounter.

You may have to abort your calculation and look for ways to simplify the integrand (e.g., through use of cylindrical or spherical coordinates in three variables) in order to move toward getting a result.

Question... Maple did not give me a number when I evaluated a double or triple integral, but I know the answer should be numerical. What should I look for?

Answer... Check for these problem areas:

- It could be a mathematical error. Did you set up the integral correctly? Recheck your limits of integration, too.

- It could be a Maple input error. Did you type the integrand correctly? Did you input the order of integration correctly?

- The function may have variables that are not involved in the integration and that don't evaluate as numbers. For example, your input function might be written in terms of x and y, but you're doing the integration with respect to polar coordinates **theta** and **r**.

CHAPTER 19

Matrices and Vectors

Matrices

Defining Matrices

The easiest way to define a matrix in Maple is by using the **Matrix** palette. For example, to enter the matrix $\begin{pmatrix} 3 & -4 & 8 \\ -1 & 0 & 5 \end{pmatrix}$, you select the **Matrix** palette, set the number of rows to 2, set the number of columns to 3, press the **Insert Matrix** button, then type:

Keystrokes: **3** *tab* **−4** *tab* **8** *tab* **−1** *tab* **0** *tab* **5**

The picture below shows the palette and the result of using it, both before and after typing the entries.

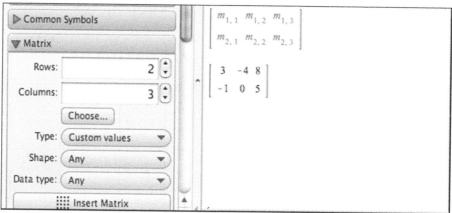

You can also use the **Matrix** command to define a matrix in the form:

Matrix([[*row₁ entries*], [*row₂ entries*], *etc.*]);

For example:

Matrix([[3,-4,8], [-1,0,5]]);

$$\begin{bmatrix} 3 & -4 & 8 \\ -1 & 0 & 5 \end{bmatrix}$$

> **Note:** The **Matrix** command needs both the [*square brackets*] around the lists and (*parentheses*) at the outer level of the command. A shortcut construction using < *angle brackets* > is discussed a little later in this chapter, see also **?<>** .

Basic Operations Involving Matrices

You can do matrix addition and subtraction, scalar multiplication and scalar addition (adding a scalar to the main diagonal), multiplication of a matrix by a matrix or vector, and exponentiation of a square matrix by using the standard operators $+$, $-$, $*$, $.$ (dot), and $\wedge$.

Consider the matrices $A = \begin{bmatrix} -1 & 0 \\ 1 & 2 \end{bmatrix}$, $B = \begin{bmatrix} 1 & 2 \\ 3 & 4 \end{bmatrix}$ and $C = \begin{bmatrix} 3 & -4 & 8 \\ -1 & 0 & 5 \end{bmatrix}$.

```
A := Matrix([[-1,0],[1,2]]):
B := Matrix([[1, 2],[3,4]]):
C := Matrix([[3, -4, 8],[-1,0,5]]):
```

Operation	Example	Maple Command
Addition (**+**)	$A + B = \begin{bmatrix} -1 & 0 \\ 1 & 2 \end{bmatrix} + \begin{bmatrix} 1 & 2 \\ 3 & 4 \end{bmatrix}$	`A + B;` $\begin{bmatrix} 0 & 2 \\ 4 & 6 \end{bmatrix}$
Subtraction (**-**)	$A - B = \begin{bmatrix} -1 & 0 \\ 1 & 2 \end{bmatrix} - \begin{bmatrix} 1 & 2 \\ 3 & 4 \end{bmatrix}$	`A - B;` $\begin{bmatrix} -2 & -2 \\ -2 & -2 \end{bmatrix}$
Scalar multiplication (*****)	$5C = 5\begin{bmatrix} 3 & -4 & 8 \\ -1 & 0 & 5 \end{bmatrix}$	`5*C;` $\begin{bmatrix} 15 & -20 & 40 \\ -5 & 0 & 25 \end{bmatrix}$
Matrix multiplication (**.**)	$AC = \begin{bmatrix} -1 & 0 \\ 1 & 2 \end{bmatrix}\begin{bmatrix} 3 & -4 & 8 \\ -1 & 0 & 5 \end{bmatrix}$	`A.C;` $\begin{bmatrix} -3 & 4 & -8 \\ 1 & -4 & 18 \end{bmatrix}$
Matrix Power (**^**)	$A^5 = \begin{bmatrix} -1 & 0 \\ 1 & 2 \end{bmatrix}^5$	`A^5;` $\begin{bmatrix} -1 & 0 \\ 11 & 32 \end{bmatrix}$
Inverse of a Matrix) (**^(-1)**)	$A^{-1} = \begin{bmatrix} -1 & 0 \\ 1 & 2 \end{bmatrix}^{-1}$	`A^(-1);` $\begin{bmatrix} -1 & 0 \\ \frac{1}{2} & \frac{1}{2} \end{bmatrix}$
Transpose of a Matrix) (**^%T**)	$A^T = \begin{bmatrix} -1 & 0 \\ 1 & 2 \end{bmatrix}^T$	`A^%T;` $\begin{bmatrix} -1 & 1 \\ 0 & 2 \end{bmatrix}$

Vectors

Column and Row Vectors

We use the same **Matrix** palette for creating vectors. If we reduce the number of rows or columns to one, the **Insert Matrix** button becomes an **Insert Vector** button. We can then create a row or column vector.

Shortcut Syntax

There is a shorthand syntax for defining matrices or vectors that uses angle brackets as delimiters. For example, a column vector can be entered as **<1, 2, 3>**, and a row vector as $< 1 \mid 2 \mid 3 >$. Note that commas (,) separate rows and vertical bars (|) separate columns. For example, a 2×3 matrix $\begin{bmatrix} 1 & 3 & 5 \\ 2 & 4 & 6 \end{bmatrix}$ can be entered as:

    `< < 1, 2 > | < 3, 4 > | < 5, 6 > >;`

or

    `< <1 | 3 | 5 >, <2 | 4 | 6 > > ;`

The standard addition, subtraction and scalar multiplication of vectors are defined using the operators **+** , **–** and *****.

For example:

```
U := <1|2|3>:    # row vector
V := <4|3|-1>:   # row vector
W := <3,2>:       # column vector

U+V,  2*U,  2*U-3*V,  4*W;
```

$$[5\ 5\ 2],\ [2\ 4\ 6],\ [-10, -5, 9],\ \begin{bmatrix} 12 \\ 8 \end{bmatrix}$$

Dot Product

The dot operator "." represents the dot product between two vectors of the same type (row or column). Using the vectors U, V, and W, that we defined earlier, we calculate

    `U.V,  W.W;`

        7, 13

Some Useful Matrix Commands

LinearAlgebra Package

The **LinearAlgebra** package includes many commands for matrix computation:

    `with(LinearAlgebra):`   #Load the package first

Operation	Maple Command and Example

CrossProduct
 	
– Find the cross product between two column vectors or two row vectors of dimension three.
	`A := < 1,2,1 >:`
	`B := < 1,1,3 >:`
	`CrossProduct( A, B );`
	
	$\begin{bmatrix} 5 \\ -2 \\ -1 \end{bmatrix}$

Determinant – Find the determinant of a square matrix.	`B := < <2,3,-2>	<5,1,1>	<1,2,0> >:` `Determinant(B);` -19

Eigenvalues
– Find the eigenvalues of a square matrix.
	`C := < <2,-1,1> | <1,0,3> | <0,1,1> >;`
	`Eigenvalues(C);`
	
	$$\begin{bmatrix} -1 \\ 2 \\ 2 \end{bmatrix}$$

Eigenvectors – Find the eigenvectors of a square matrix.	`Eigenvectors(C, output = list);` $$\left[\left[-1, 1, \left\{\begin{bmatrix} \frac{1}{4} \\ -\frac{3}{4} \\ 1 \end{bmatrix}\right\}\right], \left[2, 2, \left\{\begin{bmatrix} 1 \\ 0 \\ 1 \end{bmatrix}\right\}\right]\right]$$
	This means that the eigenvalue –1 has multiplicity 1 with eigenvector [1/4, –3/4, 1] and eigenvalue 2 has multiplicity 2 with a single eigenvector [1, 0, 1]. (C does not have a full set of eigenvectors.)

The zip and map Commands

In addition to the standard matrix and vector operations, the **map** command lets you apply operations to each entry of a matrix, and the **zip** command applies operations to corresponding entries of matrices of the same size.

```
A := Matrix( [[-1,0], [1,2]] ):
```

```
B := Matrix( [[1,2], [3,4]] ):
map( x->sqrt(x), B );        #Take the square root of each element.
```

$$\begin{bmatrix} 1 & \sqrt{2} \\ \sqrt{3} & 2 \end{bmatrix}$$

```
zip( (x,y)->x*y, A, B );  #Multiply the corresponding entries.
```

$$\begin{bmatrix} -1 & 0 \\ 3 & 8 \end{bmatrix}$$

More Examples

Matrices and Systems of Linear Equations

A major use of linear algebra is to solve systems of linear equations. You convert the system of equations into an augmented matrix, apply row operations to reduce the matrix, then convert back to a system of equations. The conversion to a matrix is done with the **GenerateMatrix** command in the **LinearAlgebra** package.

```
eqns := [2*x +y-z= 1, x+3*z = 4, -5*x-3*y+z = 2]:
vars := [x, y, z]:
```

```
with(LinearAlgebra):
A1 := GenerateMatrix(eqns, vars, augmented=true);
```

$$\begin{bmatrix} 2 & 1 & -1 & 1 \\ 1 & 0 & 3 & 4 \\ -5 & -3 & 1 & 2 \end{bmatrix}$$

Once we have converted the system of equations to a matrix, we can find the reduced row echelon form by using Gauss-Jordan elimination.

A2 := ReducedRowEchelonForm(A1);

$$\begin{bmatrix} 1 & 0 & 0 & \dfrac{23}{5} \\ 0 & 1 & 0 & -\dfrac{42}{5} \\ 0 & 0 & 1 & -\dfrac{1}{5} \end{bmatrix}$$

We can then use the **GenerateEquations** command to convert the matrix back to a system of equations and see the answer.

GenerateEquations(A2, vars);

$$\left[x = \frac{23}{5},\ y = -\frac{42}{5},\ z = -\frac{1}{5} \right]$$

The LinearSolve Command

The **LinearSolve** command can be used to solve linear systems without going through the explicit reduction of the matrix to row echelon form. For example using the augmented matrix **A1** we defined above,

```
A := A1[..,1..3]:        #A is the first three columns of A1.
b := A1[..,4]:           #b is the fourth column of A1.
with(LinearAlgebra):
LinearSolve( A, b );
```

$$\begin{bmatrix} \dfrac{23}{5} \\ -\dfrac{42}{5} \\ -\dfrac{1}{5} \end{bmatrix}$$

Vector Space operations

Another standard task in linear algebra is to find a basis of a vector space. Maple has commands to find the basis of a vector space and the basis of the intersection of two vector spaces. Let us define the vectors,

```
V1 := <0, 1, 0> :    V2 := <-2, 1, 0>:
V3 := <2, 1, 0> :    V4 := <0, 0, 1>:
V5 := <0, 1, 3> :
```

We use the **Basis** command to find a basis of a vector space that is the span of a set or list of vectors.

```
with(LinearAlgebra):
Basis( [ V1, V2, V3 ] );
```

$$\left[\begin{bmatrix} 0 \\ 1 \\ 0 \end{bmatrix}, \begin{bmatrix} -2 \\ 1 \\ 0 \end{bmatrix} \right]$$

This gives a basis of the vector space that is the span of the vectors *V*1, *V*2 and *V*3.

To find the basis of the intersection of several vector spaces, we use the **IntersectionBasis** command.

IntersectionBasis([[V2, V3], [V4, V5]]);

$$\left[\begin{bmatrix} 0 \\ 1 \\ 0 \end{bmatrix} \right]$$

This means that the vector [0, 1, 0] is the basis of the subspace which is the intersection of the two vector spaces Span(V2, V3) and Span(V4, V5).

Note that the basis vector of the intersection need not be a member of the basis of either of the spaces whose intersection is being computed.

The Student [LinearAlgebra] package

The **Student[LinearAlgebra]** package is an alternative package that can be used for linear algebra operations. It includes many commands for visualization and can help students develop their geometric understanding of linear algebra.

For example, the **ProjectionPlot** command shows the projection of one vector onto another; the **LinearTransformPlot** command shows the action of a linear transformation on the unit circle; the **EigenPlot** command shows what a linear map does to a collection of unit vectors; You can find out more about them from **?ProjectionPlot; ?LinearTransformPlot;** and **?EigenPlot;**.

When using these commands, setting the **infolevel** to **1** will give you more information. For example:

```
with(Student[LinearAlgebra]):
infolevel[Student[LinearAlgebra]] := 1:

ProjectionPlot(<3, 4>, <1, 2>);
```

```
Vector:      <3, 4>
Projection: <2.200, 4.400>
Orthogonal complement: <.8000, -.4000>
Norm of orthogonal complement: .8944
```

The Projection of a Vector
Onto a Line

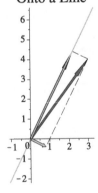

Useful Tips

 The **LinearAlgebra** package has many more useful commands, including:

- For extracting pieces of a matrix: **Row**, **Column**, **DeleteRow**, **DeleteColumn**, **Submatrix**, and **ArrayTools[Concatenate]**.
- For working with vectors: **Dimension**, **GramSchmidt**, **Normalize**, and **VectorAngle**.
- For constructing special matrices (and vectors): **BandMatrix**, **ConstantMatrix**, **DiagonalMatrix**, **IdentityMatrix**, **RandomMatrix**, **RandomVector**, **ScalarMatrix**, **ScalarVector**, **UnitVector**, **ZeroMatrix**, and **ZeroVector** (in **LinearAlgebra**).

 Some of the commands in the **Student[LinearAlgebra]** package include:

- For step-by-step row reduction: **GaussJordanEliminationTutor**.
- For understanding the geometry of a linear system: **ColumnSpace**, **LinearSystemPlot**, **NullSpace**, **Rank**, and **RowSpace**.

You should be aware that some commands are included in both the **LinearAlgebra** and the **Student[LinearAlgebra]** packages. Examples include: **Eigenvalues**, **Eigenvectors**, **CharacteristicPolynomial**, **JordanForm**, **Determinant**, and **Trace**. The **Student[LinearAlgebra]** versions are generally identical to the ones found in **LinearAlgebra**.

Troubleshooting Q & A

Question... When I used one of the commands discussed in this chapter, Maple returned the command unevaluated. What should I check for?

Answer... Check that you load the corresponding package

```
with(LinearAlgebra):
with(Student[LinearAlgebra]):
```

Question... When I tried to create a matrix with the **Matrix** command, the result had extra zeroes that I did not enter. How did this happen?

Answer... Check that the rows or columns all have the same number of entries. The **Matrix** command pads short rows out with zeroes to give a rectangular array.

Question... When I tried to add, subtract, or multiply two matrices, I got an error message. What should I check for?

Answer... Check that the matrices in the command have compatible dimensions.

CHAPTER 20

Parametric Curves and Surfaces in Space

Parametric Curves in Space

The spacecurve Command

You use the **spacecurve** command, defined in the **plots** package, to create a 3-D plot of a parametric curve in space. To see the curve given as $(x(t), y(t), z(t))$, for $a \leq t \leq b$, type:

```
with(plots):
spacecurve( [ x(t), y(t), z(t), t = a..b ] );
```

This format is similar to the syntax of the **plot** command used for plane parametric curves. Here, however, the curve is defined with *three* parametric functions rather than *two*.

■ **Example.** The helix given parametrically by $(t, 3\cos(t), 3\sin(t))$, for $0 \leq t \leq 8\pi$, is drawn with:

```
with(plots):
spacecurve( [t, 3*cos(t), 3*sin(t), t = 0..8*Pi] );
```

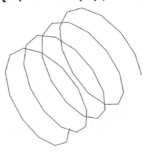

Options for spacecurve

Most options that you can use with **plot3d**, including **axes**, **labels**, **color**, and **scaling**, can also be used with **spacecurve**. The **numpoints** option provides some control of the resolution of the picture Maple produces. Here's a nicer picture of the same helix.

```
spacecurve( [t, 3*cos(t), 3*sin(t), t = 0..8*Pi],
            axes=normal,color=black, thickness=3,
            numpoints = 100, scaling=constrained);
```

Parametric Surfaces in Space

The plot3d Command

The **plot3d** command that we used to draw the graph of a two-variable function in Chapter 16 can also be used to draw a parametric surface in space. If a surface is defined parametrically by $((x(u,v), y(u,v), z(u,v))$ for $u_0 \le u \le u_1$ and $v_0 \le v \le v_1$, you enter:

```
plot3d( [  x(u, v)  ,  y(u, v)  ,  z(u, v) ],
        u =  u_0 .. u_1,  v =  v_0 .. v_1 );
```

■ **Example.** To see a portion of the one-sheeted hyperboloid given parametrically by $(\cos(u)\cosh(v), \sin(u)\cosh(v), \sinh(v))$, for $0 \le u \le 2\pi$ and $-2 \le v \le 2$, you write:

```
H:=[cos(u)*cosh(v),sin(u)*cosh(v),sinh(v)]:
plot3d( H, u = 0..2*Pi, v = -2..2);
```

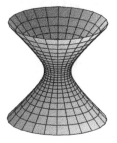

The surface $(v(2 - \cos(4u))\cos(u), v(2 - \cos(4u))\sin(u), v^2)$, for $0 \le u \le 2\pi$ and $0 \le v \le 2$, gives a very nice picture of a vase:

```
V:=[v*(2-cos(4*u))*cos(u),v*(2-cos(4*u))*sin(u),v^2]:
plot3d( V, u = 0..2*Pi, v = 0..2, grid = [60, 30] );
```

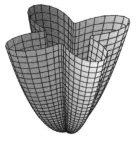

In Chapter 16 we looked at the surface $z = \dfrac{x^2 - y^2}{(x^2 + y^2)^2}$ noting that in cylindrical

coordinates it simplified to $z = \dfrac{\cos(2\theta)}{r^2}$. In that chapter we solved for r as two

functions of z and θ. The nicer alternative is to plot the surface parametrically, so
that we can describe z as a function of r and θ, and describe the surface in

parametric form as $(r\cos\theta, \, r\sin\theta, \, \dfrac{\cos(2\theta)}{r^2})$, for $0 \le \theta \le 2\pi$ and $r \ne 0$.

```
S := [ r*cos(theta), r*sin(theta),cos(2*theta)/r^2 ]:
plot3d( S, theta = 0..2*Pi, r = -3..3,
        view = -2..2, axes = boxed );
```

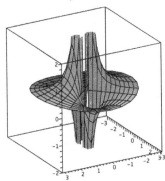

Plotting Multiple Curves and Surfaces

The **spacecurve** (or **plot3d**) command can sketch several curves (or surfaces)
with a single command, just as you've already seen in the **plot**, **polarplot**, and
implicitplot commands. Use one of these formats:

> **spacecurve({[**curve₁**, t=**a_1**..**b_1**], [**curve₂**, t=**a_2**..**b_2**]});**

> **plot3d([[**surface₁**], [**surface₂**]], u = **u_0**..**u_1**, v = **v_0**..**v_1**);**

> **Note:** Please notice that we need to use **{ curly braces }** instead of
> **[square brackets]** to group the curves together in the **spacecurve**
> command.

**Multiple
Curves**

■ **Example.** Consider the helixes $(t, -3\sin(t), 3\cos(t))$ with $\pi \le t \le 6\pi$ and
$(t, 3\cos(t), 3\sin(t))$ with $0 \le t \le 8\pi$. They can be drawn together using:

```
spacecurve( [[t,-3*sin(t), 3*cos(t),t=Pi..6*Pi],
    [t,3*cos(t), 3*sin(t),t=0..8*Pi] ],
            numpoints=100, color=black );
```

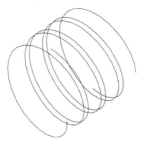

Multiple Surfaces

■ **Example.** The paraboloid $(r\cos(t),\, r\sin(t),\, 2-r^2)$ opens down, while the paraboloid $(r\cos(t),\, r\sin(t),\, r^2)$ opens up. We can combine them to form a nice "beehive."

```
plot3d( [ [r*cos(t),r*sin(t),r^2],
          [r*cos(t),r*sin(t), 2-r^2] ],
        r = 0..1, t = 0..2*Pi);
```

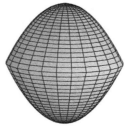

Shading, Coloring, and Transparency

Shading for Surfaces

The **plot3d** command colors the surface according to a default color and shading scheme. The goal is usually to use the color and shading (and lighting and more) to emphasize certain features of the surface.

You can turn off Maple's default shading by setting **shading=none**. This results in a plain white surface. For example:

```
plot3d( [2*cos(t), 2*sin(t), z],
        t=0..2*Pi, z=-4..4, shading=none );
```

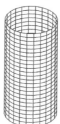

You can control the shading more directly by specifying, for example, **shading = zgrayscale** (colors lower points darker and higher points lighter) or **shading = zhue** (colors the lowest points blue and the highest points red). For example:

```
plot3d( [2*cos(t),2*sin(t),z],t = 0..2*Pi, z=-4..4,
        shading=zgrayscale, scaling=constrained );
```

Coloring Function

You can explicitly set the coloring of points on a surface by using the **color** option. If the color is a list of three numbers, this is interpreted as an **RGB** color. For example,

```
s := (u,v) -> [u^2, v, v^3];
plot3d( s(u,v), u=-2..2, v=-2..2,
        style=patch,
        color=[(u+v)/4, u^2/4, sin(v)^2]);
```

This means that each point $s(u,v)$ is painted with the color defined by

COLOR$(\mathrm{RGB}, \dfrac{u+v}{4}, \dfrac{u^2}{4}, \sin^2(v)\,)$. For example, at the point $s(0,1)=[0,1,1]$ the color is **COLOR**$(\mathrm{RGB}, 1/4, 0, \sin(1)\hat{}2)$, a dark blue.

Transparency

When we plot several surfaces together we may want one surface to be partly transparent so that we can see another surface behind it. This can be done by setting the **transparency** option to be a number between 0 and 1. For example the following puts together a sphere and a cylinder, where we can see the cylinder through the sphere:

```
with(plots):
sphere := plot3d( 2, phi=0..Pi, theta=0..2*Pi,
                coords = spherical,
                color=red, transparency=.5 ):
cylinder := plot3d( 1, theta=0..2*Pi, z=-2..3,
                coords=cylindrical,
                color=green ):
display( [ sphere, cylinder ] );
```

(The figure is not shown here, since color transparency does not show up well in a static black and white image. Try it for yourself!!)

More Examples

Combining Graphics

You can combine curves, surfaces, and other three-dimensional images into a single graphic using the **display** command (just as we saw in Chapters 10 and 16).

■ **Example.** The upper hemisphere of the unit sphere $x^2 + y^2 + z^2 = 1$ is best plotted in spherical coordinates with $\rho = 1$, for $0 \le \theta \le 2\pi$ and $0 \le \phi \le \pi / 2$,

```
g1:=plot3d( 1, theta=0..2*Pi, phi=0..Pi/2,
            coords=spherical):
```

The point $P = (\frac{1}{2}, \frac{1}{2}, \frac{1}{\sqrt{2}})$ lies on this hemisphere. A normal (perpendicular) line to the hemisphere at P is given by $\vec{r}(t) = (\frac{1}{2}+t, \frac{1}{2}+t, \frac{1}{\sqrt{2}}+\sqrt{2}t)$. This command shows just a portion of the normal line:

```
with(plots):
g2:=spacecurve( [1/2+t, 1/2+t, 1/sqrt(2)+sqrt(2)*t ],
                t=0..0.15, thickness=3, color=blue ):
```

Finally, the plane tangent to the hemisphere at P has equation $z = (2 - x - y) / \sqrt{2}$. You can sketch a portion of it near the point P with:

```
g3:=plot3d( (2-x-y)/sqrt(2),
            x = 0.2..0.8, y = 0.2..0.8,
            grid=[2,2], color=cyan):
```

You can now see one of the nicest features of Maple, the ability to combine these graphics, despite the fact that each was drawn using a different type of command.

```
display( [g1,g2,g3], orientation=[97,78],
         scaling=constrained );
```

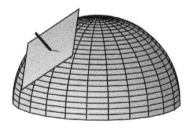

Notice how well the graphic convinces you that we have described both the tangent plane and the normal line correctly.

Troubleshooting Q & A

Question... When I tried to use the **spacecurve** or **display** command, Maple returned the input unevaluated. What went wrong?

Answer... You forgot to load the **plots** package first. Type:

```
with(plots);
```

Question... When I tried to draw a parametric curve or surface in space, I got an error message. What should I check?

Answer... There are many possibilities, but our best suggestions are:

- Did you use the **spacecurve** or **plot3d** command? A common mistake is trying to draw a 3-D picture with the **plot** command.
- Check that you have followed the correct syntax of entering the commands. The formats for **spacecurve** and **plot3d** are different and can cause confusion.
- Did you enter the function(s) correctly without a typing mistake? Is each of the functions defined everywhere in the interval(s) you specified?
- Did you use the same literal parameter in both your function and interval? (For example, check that you didn't write **f(x,y)** when the parameters were **u** and **v**.)
- Did you use two parameters for a surface? One variable for a curve?

Question... When I used **plot3d** to draw *one* parametric surface, I got *three* surfaces instead. What happened?

Answer... On some rare occasions, it can happen that your parametric surface is actually split into three pieces. However, most likely, you entered the command incorrectly. A common mistake is to type **{** *curly braces* **}** instead of **[** *square brackets* **]** for the coordinate functions. Maple will then interpret your input as a request to draw three *graphs* (see Chapter 16).

CHAPTER 21
Vector Fields

Drawing a Vector Field

The fieldplot Command

The **fieldplot** command (defined in the **plots** package) sketches vector fields in the plane. To see the vector field defined by $\vec{F}(x, y) = (F_1(x, y), F_2(x, y))$ for $x_0 \le x \le x_1$, $y_0 \le y \le y_1$, you enter:

```
with( plots ):
fieldplot( < F₁(x, y),  F₂(x, y) >,  x = x₀ .. x₁,  y = y₀ .. y₁ );
```

For example, to see the vector field $(-y, x)$, for $-2 \le x \le 2$, $-2 \le y \le 2$, you type:

```
with( plots ):
fieldplot( <-y,x>, x=-2..2, y=-2..2);
```

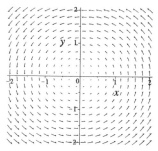

Some Useful Options

Some options for the **fieldplot** command can help us to customize the plot.

Option	What It Does
grid=[10,10]	Draws 10 × 10 = 100 vectors. The default value is **[20, 20]**.
arrows=thick	Uses thicker arrows for vectors.
color=blue	The arrows will be drawn in blue (or whatever color you specify).

Strictly speaking, Maple does not draw the vector field $(-y, x)$. In an attempt to avoid overlapping arrows that obscure information about the field, Maple uses a combination of length and thickness of the arrows to indicate the length of the vector. This is most effective when the **arrows=thick** option is used as illustrated in the next picture.

```
fieldplot( <-y,x>, x=-2..2, y=-2..2, grid = [10,10],
          arrows = thick, color = grey );
```

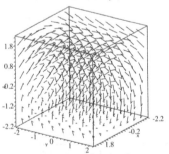

Vector Fields in Space

To draw a 3-D vector field (in space), you use the **fieldplot3d** command, also defined in the **plots** package.

To see the vector field $\vec{F}(x, y, z) = (F_1(x, y, z),\ F_2(x, y, z),\ F_3(x, y, z))$, for the intervals $x_0 \le x \le x_1$, $y_0 \le y \le y_1$, and $z_0 \le z \le z_1$, you type:

```
with(plots):
fieldplot3d( <F₁(x, y, z),  F₂(x, y, z),  F₃(x, y, z)>,
             x = x₀ ..x₁,  y = y₀ ..y₁,  z = z₀ ..z₁ );
```

■ **Example.** To see the vector field $(-z,\ 1,\ x)$ inside a cube surrounding the origin, we will use:

```
with(plots):
fieldplot3d( [-z,1,x], x=-2..2, y=-2..2, z=-2..2,
             axes = boxed);
```

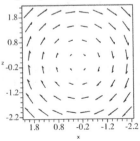

This looks like a mess. If you click on the picture and rotate the frame box, you can get a better view of the vector field by looking at it from the positive, y-axis, say with **orientation = [90, 90]**.

Vector Fields in Different Coordinate Systems

The VectorCalculus Package

The **VectorCalculus** package contains several commands for working with vector fields. However, you need to use the **VectorField** command to define the vector field you'll be using, since it encodes both your definition of the field and your choice of coordinate system.

```
with( VectorCalculus ):
VectorField(<2*x+2*y,2*x,-3*z^2>,'cartesian'[x,y,z]);
```

$$(2x + 2y)\overline{\mathbf{e}}_x + 2x\overline{\mathbf{e}}_y - 3z^2\overline{\mathbf{e}}_z$$

The symbols $\overline{\mathbf{e}}_x$, $\overline{\mathbf{e}}_y$ and $\overline{\mathbf{e}}_z$ that appear in the output above denote the standard basis elements for three-space. Since you will do most of your work with vectors in the three-dimensional Cartesian coordinate system, it makes sense to set that coordinate system as the default.

```
SetCoordinates( 'cartesian'[x,y,z] ):
VectorField( < 2*x+2*y, 2*x, -3*z^2> );
```

$$(2x + 2y)\overline{\mathbf{e}}_x + 2x\overline{\mathbf{e}}_y - 3z^2\overline{\mathbf{e}}_z$$

We can have Maple use the column format instead of the linear combination format by setting **BasisFormat** to **false**.

```
BasisFormat( false ):
VectorField( <2*x+2*y,2*x,-3*z^2> );
```

$$\begin{bmatrix} 2x + 2y \\ 2x \\ -3z^2 \end{bmatrix}$$

We will generally leave the **BasisFormat** at its default setting of *true* to remind us of the coordinate system we are using.

```
BasisFormat(true):
```

Other Coordinate Systems

Maple understands many standard coordinate systems (Cartesian, polar, cylindrical, and spherical) and some esoteric ones as well (e.g. hyperbolic, rose, and toroidal).

The **SetCoordinates** command can be used to set the default coordinate system. For example,

```
with(VectorCalculus):
SetCoordinates('cartesian'[x,y]):
SetCoordinates('cartesian'[x,y,z]):
SetCoordinates('polar'[r,theta]):
SetCoordinates('cylindrical'[r,theta, z]):
SetCoordinates('spherical'[rho,theta,phi]):
```

> **Note:** In the commands above, the order of the variables such as **[x, y, z]**, **[r, theta, z]**, and **[rho, theta, phi]** is important. If you list them in a different order, Maple will misinterpret your input.

We can also assign the coordinates used for a specified vector or vector field without changing the default coordinate system.

```
SetCoordinates('cartesian'[x,y]):              # set as default
v1 := SetCoordinates( <3,4>, 'polar'[r,theta]);
```
$$3\mathbf{e}_r + 4\mathbf{e}_\theta$$
```
v2 := VectorField( <r*theta, theta>, 'polar'[r,theta]);
```
$$(r\theta)\overline{\mathbf{e}}_r + (\theta)\overline{\mathbf{e}}_\theta$$

Here, we are able to define a vector and a vector field in terms of $\overline{\mathbf{e}}_r$ and $\overline{\mathbf{e}}_\theta$ without changing the default Cartesian coordinate system.

Change of Coordinates

To convert a vector or vector field from one coordinate system to another, you use the **MapToBasis** command, naming the system you are converting to along with the variables used.

```
SetCoordinates( 'cartesian'[x,y] ):
V1 := <3,4>;
```
$$3\mathbf{e}_x + 4\mathbf{e}_y$$
```
V2 := MapToBasis( V1,'polar'[r,theta] ):
simplify(%);
```
$$5\mathbf{e}_r + \arctan\left(\frac{4}{3}\right)\mathbf{e}_\theta$$
```
SetCoordinates( 'cartesian'[x,y,z] ):
V1 := VectorField( <x, y, z > );
```
$$(x)\overline{\mathbf{e}}_x + (y)\overline{\mathbf{e}}_y + (z)\overline{\mathbf{e}}_z$$
```
V2 := MapToBasis( V1,'cylindrical'[r,theta,z] ):
simplify(%);
```
$$(r)\overline{\mathbf{e}}_r + (z)\overline{\mathbf{e}}_z$$

Gradient, Curl, and Divergence

Gradient Field

The gradient of a function f is denoted by grad f. The gradient field of f always points in the direction of steepest increase of the value of f. In Cartesian coordinates it is defined to be the vector

$$\text{grad}\, f = f_x \vec{e}_x + f_y \vec{e}_y, \text{ or } \text{grad}\, f = f_x \vec{e}_x + f_y \vec{e}_y + f_z \vec{e}_z,$$

depending on whether f is a function of two or three variables, respectively. We find grad f with the **Gradient** function from the **VectorCalculus** package (which we assume is already loaded):

```
SetCoordinates( 'cartesian'[x,y,z] ):
f := (x,y,z) -> x^2 + 2*x*y - z^3:
Gradient( f(x,y,z) ) ;
```
$$(2x + 2y)\overline{\mathbf{e}}_x + 2x\overline{\mathbf{e}}_y - 3z^2\overline{\mathbf{e}}_z$$

The gradient is often written in terms of the "del" or "nabla" operator, which is a vector of partial derivative operators.

$$\nabla = \left(\frac{\partial}{\partial x} \quad \frac{\partial}{\partial y} \quad \frac{\partial}{\partial z} \right)$$

We can also obtain the gradient by using either the **Del** or **Nabla** commands or by using the "del" symbol from the "**Common Symbols**" palette. Thus we get the same result with any of the following

```
Del( f(x,y,z) );
Nabla( f(x,y,z) );
∇ ( f(x,y,z) );    #∇ is from the Common Symbols palette
```

Curl and Divergence

The curl and divergence of a vector field measure the rotation and spreading of particles whose motion is defined by the field, respectively. The curl and divergence of a vector field are calculated using the **Curl** and **Divergence** commands from the **VectorCalculus** package.

```
SetCoordinates( 'cartesian'[x,y,z] ):

Curl( VectorField( <x+y, y+z, sin(x+y)+z^2> ) );
```
$$(\cos(x+y)-1)\overline{e}_x - (\cos(x+y))\overline{e}_y - \overline{e}_z$$

```
Divergence( VectorField( <x^2,y^2,z^2> ) );
```
$$2x + 2y + 2z$$

The curl and divergence can also be computed by taking the cross product and dot product, respectively, of the del operator with the vector field. Since **VectorCalculus** is already loaded, this can be done with the **CrossProduct** and **DotProduct** commands or with the **.** and **&x** operators respectively.

We thus have three ways of computing these derivatives.

```
vF :=
VectorField(<F[1](x,y,z),F[2](x,y,z),F[3](x,y,z)>);
```

To calculate the curl, use any of	*To calculate the divergence, use any of*

`Curl( vF );`
`CrossProduct( Del, vF );`
`Del &x vF;`
	`Divergence( vF );`
	`DotProduct( Del, vF );`
	`Del . vF;`

Other Coordinate Systems

The **Gradient**, **Curl**, and **Divergence** commands can also be used in coordinate systems other than the Cartesian coordinate system. With **Gradient** you simply specify the coordinate system and variable list. For the **Curl** and **Divergence** commands, you don't need to mention the coordinate system, because the system is explicitly included when a vector field is defined.

Coordinate System	Maple Command
Gradient in the Cartesian system	`Gradient( x^2 + 2*x*y - z^3, [x,y,z] );` $$\left(2x+2y\right)\overline{\mathbf{e}}_x + 2x\overline{\mathbf{e}}_y - 3z^2\overline{\mathbf{e}}_z$$

Gradient in the cylindrical system	`Gradient( r*cos(theta)+r^2*z,`
	      `'cylindrical'[r,theta,z] );`
	$$\left(\cos(\theta)+2rz\right)\overline{\mathbf{e}}_r - \sin(\theta)\overline{\mathbf{e}}_\theta + \left(r^2\right)\overline{\mathbf{e}}_z$$

Gradient in the spherical system	`Gradient( 2*rho^2*cos(theta)+sin(phi),`
	      `'spherical'[rho, theta, phi] );`
	$$4\rho\cos(\theta)\overline{\mathbf{e}}_\rho - 2\rho\sin(\theta)\overline{\mathbf{e}}_\theta + \left(\frac{\cos(\phi)}{\rho\sin(\theta)}\right)\overline{\mathbf{e}}_\phi$$

Divergence in the spherical system	`F := VectorField( <rho^2, rho*sin(phi),`
	    `sin(theta)>, 'spherical'[rho,theta,phi] ):`
	`simplify( Divergence(F) );`
	$$\frac{4\rho\sin(\theta) + \sin(\phi)\cos(\theta)}{\sin(\theta)}$$

Curl in the cylindrical system	`G := VectorField( <theta, z, r>,`
	        `'cylindrical'[r,theta,z]):`
	`simplify( Curl(G) );`
	$$-\overline{\mathbf{e}}_r - \overline{\mathbf{e}}_\theta + \left(\frac{z-1}{r}\right)\overline{\mathbf{e}}_\phi$$

Line and Flux Integrals

Engineers are often interested in integrating a vector field $\vec{F}$ either along a curve $\vec{r}(t)$, for $a \le t \le b$, or over a surface $\vec{s}(u,v)$, for $u_0 \le u \le u_1$, $v_0 \le v \le v_1$. These are defined as:

- Line integral: $\displaystyle\int_C \vec{F}\cdot ds = \int_a^b \vec{F}(\vec{r}(t))\cdot\vec{r}\,'(t)dt$

- Flux integral: $\displaystyle\iint_S \vec{F}\cdot\vec{n}\,dS = \pm\int_{u_0}^{u_1}\int_{v_0}^{v_1}\vec{F}(\vec{s}(u,v))\cdot\left(\frac{\partial\vec{s}}{\partial u}\times\frac{\partial\vec{s}}{\partial v}\right)dv\,du$ (The choice of

the $\pm$ sign depends on how the normal vector for the surface is defined.)

These integrals are easily computed using the **LineInt** and **Flux** commands defined in the **VectorCalculus** package as we show in the following examples.

Line Integral

■ **Example.** Let $\vec{F}$ be the vector field $\vec{F}(x, y) = (x + y, -y)$ and $\vec{r}(t)$ the parametric curve $\vec{r}(t) = (1-t, t^2)$ for $0 \le t \le 1$. To compute the line integral, first we set up the field and the path.

```
SetCoordinates( 'cartesian'[x,y] ):
F := VectorField( <x+y, -y> ):    # defines the vector field F⃗.
r := < 1-t, t^2 >:                # defines the curve r⃗(t).
C := Path( r, t=0..1 ):           # defines the path C.
```

We can compute the integrand $\vec{F}(\vec{r}(t))$ by using the **evalVF** command to evaluate **F** on **r**, and then take the dot product of that with the derivative of **r**.

```
DotProduct( evalVF(F,r), diff(r,t) );
```

$$-1+t-t^2-2t^3$$

The line integral can then be found with

```
int( %, t=0..1 );
```

$$-\frac{4}{3}$$

Or we can use the **LineInt** command from the **VectorCalculus** package to compute the line integral directly:

```
LineInt( F, C );
```

$$-\frac{4}{3}$$

Flux Integral

■ **Example.** Suppose the flux integral of $\vec{F}(x, y, z) = (x(1+z), y, 0)$ over the surface S is given by

$$\int_1^4 \int_0^{2\pi} \vec{F}(\vec{s}(u,v)) \cdot \left(\frac{\partial \vec{s}}{\partial u} \times \frac{\partial \vec{s}}{\partial v} \right) dv\,du$$

where S is parameterized by $\vec{s}(u,v) = (u\cos v, u\sin v, u)$. We can evaluate this integral as follows, noting that we use the **Surface** command to describe the parametric surface. First we set things up:

```
SetCoordinates( 'cartesian'[x,y,z] ):
F := VectorField( <x*(1+z),y,0> ):
s := (u, v) -> < u*cos(v), u*sin(v), u >:
S := Surface(s(u,v), [u,v]=Rectangle(1..4,0..2*Pi)):
```

The integrand can be found using explicit commands for the dot and cross product:

```
DotProduct( evalVF(F, s(u,v) ),
            CrossProduct( diff(s(u,v), u),
                          diff(s(u,v), v) ) );
```

$$-u^2 \cos(v)^2 (1+u) - u^2 \sin(v)^2$$

Then we can compute the flux integral using iterated **int** commands or **MultiInt**:

```
int( int( q, u=1..4 ), v=0..2*Pi );
```

$$-\frac{423\pi}{4}$$

```
with( Student[MultivariateCalculus]):
```

```
MultiInt( q, u=1..4, v=0..2*Pi );
```

$$-\frac{423\pi}{4}$$

Alternatively, the **Flux** command can be used to evaluate the flux integral directly:

```
Flux( F, S );
```

$$-\frac{423\pi}{4}$$

More Examples

**The Student[
VectorCalculus]
Package**

The **Student[VectorCalculus]** package contains all the commands in the VectorCalculus package that we mentioned in this chapter. In addition, this package also provides more graphic options to help the students develop their understanding of vector calculus. You can load this package with:

```
with(Student[VectorCalculus]):
```

For example, when defining a vector field with this package loaded, you can obtain a plot of the vector field by adding the option **output=plot**. The **view** option can be added if you want to specify the viewing window.

```
VectorField( <-y, x>,
             output=plot, view=[-2..2, -2..2]);
```

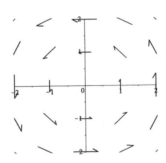

In an earlier example computing a line integral, we used:

```
SetCoordinates( 'cartesian'[x,y] ):

F := VectorField( <x+y, -y> ):
C := Path( <1-t,t^2>, t=0..1 ):

LineInt( F, C );
```

$$-\frac{4}{3}$$

Now, using the **LineInt** command inside the **Student[VectorCalculus]** package, we can add the **output=plot** option and obtain:

```
LineInt( F, C, output=plot );
```

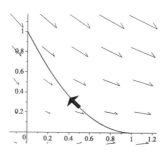

As you can see from the picture, if a particle moves along the curve from its starting point (1,0) to its ending point (0,1), it is moving against the force of the vector field. Thus, the line integral is negative.

The Perpendicular Property of the Gradient

■ **Example**. Consider the function $f(x, y) = xy + 2x$:

```
f := (x,y) -> x*y +2*x:
```

The perpendicular property of the gradient vector states that grad $f(a,b)$ is perpendicular to the level curve of f that goes through (a,b). To see this, we create a plot of the gradient field of f:

```
with( Student[VectorCalculus] ):
SetCoordinates( 'cartesian'[x,y] ):
pict1 := VectorField( Gradient(f(x,y)),
          output=plot, view = [-4..4, -4..4],
          fieldoptions = [grid=[10,10], arrows=thick]):
```

Next we need a contour plot using the command from the **plots** package:

```
with( plots ):
pict2 := contourplot( f(x,y), x=-4..4, y=-4..4,
                      contours=20, color=blue):
```

Now combine the contours and gradient vectors into one picture:

```
display([pict1, pict2]);
```

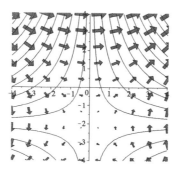

Do you agree that the gradient vectors are perpendicular to the level curves at every point?

Useful Tips

☼ ☼   In many cases, you can draw a less cluttered (and faster) picture of vector fields by using a smaller number of gridlines, such as **grid =[10,10]**. (Default grid is **[20,20]**).

☼ ☼   Both the **VectorCalculus** and **Student[VectorCalculus]** packages have commands for a number of other "derivatives" typically used in the subject. These include **Hessian**, **Jacobian**, and **Laplacian.** Check the appropriate help pages for more details. If a vector field is the gradient of a potential function, the **ScalarPotential** command can be used to recover the potential function. Similarly, if a vector field is the curl of a potential field, the **VectorPotential** command can be used to recover the potential field.

☼ ☼ ☼   Instead of using a parametric **Path** to describe a given curve for the **LineInt** command, you may be able to define the curve more directly using one of the **Circle**, **Ellipse**, **Line**, **LineSegments**, or **Arc** commands. For example:

```
F := VectorField( <x+y, -y>, 'cartesian'[x,y] ):
LineInt( F, Circle(<3,4>,2) );   # circle: center (3,4) radius 2
LineInt( F, Line(<1,1>, <2,5>) );    # line from (1,1) to (2,5)
```

☼ ☼ ☼   Similarly, with the **Flux** command, you can use **Sphere** or **Box** to describe a surface without giving an explicit parameterization.

For example, given the vector field

```
SetCoordinates( 'cartesian'[x,y,z] ):
F := VectorField( <x+y, -y, z> ):
```

we can compute its flux integral along the six boundary surfaces of the box $1 \le x \le 2$, $3 \le y \le 4$, and $5 \le z \le 6$ with outward pointing normal.

```
Flux( F, Box(1..2, 3..4, 5..6, outward) );
```

To compute the flux along the sphere of center at (1,1,1) radius 3 with inward pointing normal, we use:

```
Flux( F, Sphere( <1,1,1>, 3, inward) );
```

Troubleshooting Q & A

Question...   When defining a vector field, I got an error that the first argument was the wrong type. What should I check?

Answer...   For example, a common mistake is to enter the vector field $(x, x+y)$ as **(x, x+y)** instead of either **[x, x+y]** or **<x, x+y>**.

Question... I got an error message that something was unsuitable for the current coordinate system. What was my mistake?

Answer... You've likely switched between coordinate systems without realizing it. Most commonly this is done when shifting between 2 and 3 variables in Cartesian systems. Use **SetCoordinates** to declare the appropriate coordinate system.

Question... I see the del symbol (∇) in the **Common Symbols** palette, but Maple won't let me click on it. What do I have to do to be able to activate this icon?

Answer... This icon, and several others in the **Common Symbols** palette, are inactive if you are using **Maple Input**. To activate these icons, change the input style to 2D Input ($\boxed{control -}$ R is one shortcut). Note that these symbols are active within Text regions as well.

Question... When I used **Gradient**, **Curl**, or **Divergence**, Maple returned my input unevaluated. What was my mistake?

Answer... You forgot to load the **VectorCalculus** or **Student[VectorCalculus]** package before you used these commands. Correct the problem by loading the package with one of the following:

```
with( VectorCalculus ):
with( Student[VectorCalculus] ):
```

Question... Why would I get a wrong answer when using **Gradient**, **Curl**, or **Divergence**?

Answer... There are several possibilities:

- Check that the input syntax is correct.
- You have to type the vectors **[x, y, z]**, or **[r, theta, z]**, or **[rho, theta, phi]**, either in setting the coordinate system or inside these commands, with the variables exactly in this order. If you list them in a different order, the computation will be incorrect.
- The input function or vector field has to be expressed in terms of the variables of your chosen coordinates. For example, if you want to calculate the gradient of $x + y + z$ using cylindrical coordinates, you have to enter the function as **r*cos(theta)+r*sin(theta)+z**.

Question... Why does the **Flux** command give an error message about an unknown region of integration?

Answer... You probably misspelled the type of region. Check the syntax by looking at the help pages for **Flux** and **int** (in the **VectorCalculus** package).

CHAPTER 22

Basic Statistics on a Data Sample

Numerical Measures of Data

Mean, Median, Standard Deviation, etc.

Standard statistical measures of data distribution are available in Maple. To use them, you first have to load the **Statistics** package:

```
with(Statistics):
```

■ **Example.** Let us consider the number of home runs for the 30 baseball teams in the major leagues for the 2007 season.

```
homeRuns := [201, 213, 177, 166, 171, 123, 179, 178, 176,
177, 231, 153, 201, 204, 187, 142, 165, 151, 171, 171, 129,
141, 148, 167, 118, 171, 102, 190, 131, 123];
```

We can find the mean and median directly with the following commands.

```
Mean( homeRuns );
```
$$165.2333333$$

```
Median( homeRuns );
```
$$171.$$

Other measures of central tendency such as the **Mode**, **HarmonicMean**, and **GeometricMean** are also available and work in the same way.

The best-known measures of data variability also have their customary names:

```
Variance(homeRuns);
```
$$920.5298853$$

```
StandardDeviation(homeRuns);
```
$$30.34023542$$

We can also get a summary overview of the data sample with a single command:

```
DataSummary(homeRuns);
```
[*mean* = 165.2333333, *standarddeviation* = 30.34023542, *skewness* = –0.06725916525, *kurtosis* = 2.497667841, *minimum* = 102., *maximum* = 231., *cumulativeweight* = 30.]

It should be noted that Maple calculated the sample variance and standard deviation as $\dfrac{1}{N-1}\sum_{i=1}^{N}(X_i - \bar{X})^2$ and $\sqrt{\dfrac{1}{N-1}\sum_{i=1}^{N}(X_i - \bar{X})^2}$ respectively.

Sorting and Tallying a Sample

The **Statistics** package has a **Sort** command that lets you sort the values in a sample and arrange them in increasing order.

> **homeRunsSorted := Sort(homeRuns);**
>
> [102, 118, 123, 123, 129, 131, 141, 142, 148, 151, 153, 165, 166, 167, 171, 171, 171, 171,176, 177, 177, 178, 179, 187, 190, 201, 201, 204, 213, 231]

The other standard manipulation of a data set is to count how many times each value occurs. This is done with the **Tally** command defined in the **Statistics** package. For example suppose we ask five people on the street to choose red or blue and record their answers in **colorList**.

> **colorList := [red, red, blue, blue, red];**

We can now count the answers with the **Tally** command:

> **Tally(colorList);**
>
> [*red* = 3, *blue* = 2]

Importing Numerical Values from a File

You can easily import data values into Maple from another application (e.g., a spreadsheet or an e-mail message) and then use the statistical tools we've introduced in this chapter to study the data.

For example, suppose you have a data file named "homeRun.xls" where individual data values are written in an Excel file format. You can import all the data values with the **ImportData** command.

> **homeRunList := ImportData();**

You can also import data by selecting **Import Data** from the list of **Assistants** in the **Tools** menu.

Visualizing a Data Set

Histogram

Given a set of data, we most easily visualize it with the **Histogram** command in the **Statistics** package. We see the home run data from above visually with

> **Histogram(homeRuns);**

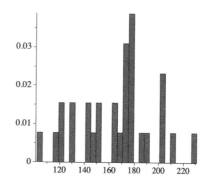

We typically find it useful to graph the frequency in absolute numbers rather than in percentages of the collection. Also, when visualizing data, we may want to sort the data into bins of fixed width. (Think of grades where 90 to 100 all count as an A.) We can specify these with the corresponding options in the **Histogram** command.

```
Histogram( homeRuns, frequencyscale=absolute,
           binwidth=10, range=49.5..199.5);
```

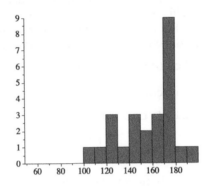

Another standard visualization of data is produced by the **PieChart** command that's also defined in the **Statistics** package.

```
PieChart(homeRuns, color=white..gray);
```

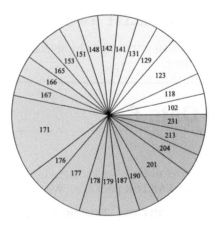

Probability Distributions

The **Statistics** package contains definitions of common probability distributions. These include the following, along with their usual parameter specifications:

Normal(μ, σ)	Binomial(n, p)	Uniform(a, b)
StudentT(ν)	Exponential(α, a)	Beta(α, β)
Poisson(λ)	DiscreteUniform(a, b)	ChiSquare(n)

Cumulative Distribution and Probability Density Functions

You can work with the *cumulative distribution function* and the *probability density function* (or *probability function*) of each of these distributions using the commands:

```
with(Statistics):
CDF(distribution, x);        # Cumulative distribution function
PDF(distribution, x);        # Probability distribution function
```

■ **Example.** Consider the normal distribution with mean 10 and deviation 3.

```
with(Statistics):
distribution1 := Normal(10,3):
```

To find the values of its cumulative distribution function and probability density function, say at $x = 15$, use:

```
CDF( distribution1, 15.0 );
```

$$0.952209647727185304$$

```
PDF( distribution1, 15.0 );
```

$$0.03315904624$$

Also, we can see the graphs of the cumulative distribution and probability density functions, respectively, with:

```
plot([ PDF( distribution1, x ),
       CDF( distribution1, x ) ],
     x = 0..20, legend=["PDF", "CDF"]);
```

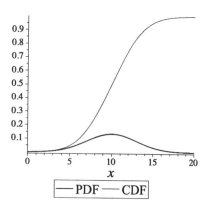

■ **Example.** We can find the cumulative distribution function of a discrete distribution in a similar way. For example, consider the binomial distribution:

```
Bdist := Binomial( 15, 0.7 ):
```

To find the value of its cumulative distribution function at 10:

```
CDF( Bdist, 10 );
```

$$0.4845089408$$

However, when working with a discrete distribution, we need to use **ProbabilityFunction** instead of **PDF**. For example, we can plot the "graph" of the probability function of the previous binomial distribution with:

```
data := [seq([i, ProbabilityFunction(Bdist, i)],
         i = 0 .. 15)]:
```

```
plot(data, style = point, symbol = solidcircle,
     title = "Binomial Probs, n=15, p=.7");
```

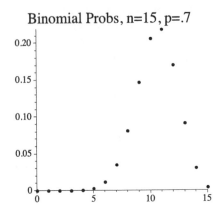

Binomial Probs, n=15, p=.7

More Examples

IQ Score

■ **Example**. It is generally believed that IQ scores are normally distributed with mean 100 and standard deviation 15. Let us theoretically estimate how many people in the world have an IQ higher than 180.

The probability that a person's IQ is between 0 and 180 is:

```
evalf( CDF( Normal(100, 15), 180) );
```

$$0.9999999518$$

So the probability that a person's IQ is higher than 180 is:

```
GR := 1 - %;
```

$$4.82 \ 10^{-8}$$

There are almost 6.75 billion people in this world, so:

```
GR * 6.75*10^9;
```

$$325.3500000$$

Theoretically, there are barely 325 people in the world having IQs higher than 180. Isn't it odd, then, how often we meet people who claim to have such high IQs?

Useful Tips

    The **ImportData** command can be used to read large datasets into a Maple array. In such situations Maple sometimes displays a placeholder instead of the actual data. For example,

```
M := Matrix( 16, 16, (i,j)->1/(i+j-1) );
```

$$\begin{bmatrix} \textit{16x16 Matrix} \\ \textit{Data Type : anything} \\ \textit{Storage : rectangular} \\ \textit{Order : Fortran\_order} \end{bmatrix}$$

Maple decides to produce this response when the matrix has either more rows or more columns than the current value of the **rtablesize** parameter. When this happens, you can still use the **Matrix Browser** to view the matrix. To activate the **Matrix Browser**, select **Browse** from the context menu produced when you right-click anywhere on the above output. The value of the **rtablesize** parameter can be changed with the **interface** command:

```
interface( rtablesize=50 );
```

$$10$$

The output from this **interface** command means that the previous setting of **rtablesize** was 10. Now, Maple will show the actual contents of any matrix (or array) that does not have more than 50 rows or 50 columns.

```
M;
```

(Maple will now show you a 16×16 matrix, but it's too big to display in here.)

Earlier, we saw how to use the **style** option in the **plot** command to graph the probability function for a discrete distribution. Other ways to present a set of numerical data in a graph are discussed in Chapter 23.

You can also use **SurvivalFunction** and **InverseSurvivalFunction** inside the **Statistics** package to compute the *survival function* and *inverse survival function* of a distribution.

CHAPTER 23
Curve Fitting

When trying to fit a curve to a set of data points, there are two traditional approaches. You can either try to find a curve that hits the points exactly (**interpolation** or **spline**) or find a curve of specified degree that gives a good approximation of the data (**regression**). Maple lets you use both approaches.

Graphical Presentation of Data

Scatter Diagrams

A set of numerical data can sometimes be presented graphically by using a scatter diagram. Depending on the type of data, you can either use **pointplot** (defined in the **plots** package) or **ScatterPlot** (defined in the **Statistics** package) to see the diagram.

```
with(plots);
pointplot( list of ordered pairs );

with(Statistics):
ScatterPlot( list of x-coordinates, list of y-coordinates );
```

> **Note:** You use the **pointplot** command if the data is a list of pairs. You use the **ScatterPlot** command if you have separate lists of *x*- and *y*-coordinates.

For example,

```
with(plots):
pointplot( [[1, 3], [2, 1], [6,2], [4, -1]],
           symbol = solidcircle);
```

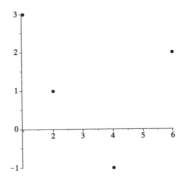

This picture an also be drawn with the **ScatterPlot** command.

```
with(Statistics):
ScatterPlot( [1,2,6,4], [3,1,2,-1],
             symbol = solidcircle);
```

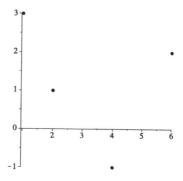

Regression

Using a Least Squares Fit

When given a data set with two variables, we often want to assume that one of the variables ought to be a function of the other. In the simplest case we may want to find the line that "best" fits this data set. (Best fit is usually determined by the "least squares method.") This process is called **linear regression**. We can add the best fitting line to the **ScatterPlot** by using the **fit** option.

```
with(Statistics):
ScatterPlot( list of x-coordinates, list of y-coordinates
             fit =[ type of curve to fit, variable]);
```

For example, to plot the points (1, 2), (3, 4), (5, 7), and the best fit line we have:

```
ScatterPlot( ([1,3,5], [2,4,7]), fit=[a*x+b,x]);
```

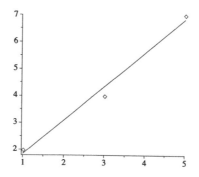

If we are only interested in the equation of the best fit line, we can use the **Fit** command from the **Statistics** package.

```
with(Statistics):
Fit(model function, list of x-coordinates, list of y-coordinates, variable);
```

```
Fit( a*x+b, [1,3,5], [2,4,7], x);
```
$$1.25000000000000000\,x + 0.583333333333333148$$

Note that the answer from the **Fit** command is expressed in terms of real coefficients, even when the input contains only integers.

■ **Example.** Here are data from the UNESCO 1990 *Demographic Yearbook* which show male life expectancy and female life expectancy in Argentina, Austria, Afghanistan, Algeria, Angola, Bolivia, Brazil, Belgium, Bangladesh, and Botswana, respectively.

```
male := [ 65.5, 73.3, 41, 61.6, 42.9,
          51, 62.3, 70, 56.9, 52.3]:
female := [ 72.7, 79.6, 42, 63.3, 46.1,
            55.4, 67.6, 76.8, 56, 59.7]:
```

Now, let us find the line that best fits the data.

```
line1 := Fit(a*x+b, male, female,x);
```
$$1.12885025120340887x - 3.19208248941262738$$

To see how well the line fits the data, we first draw the points using **ScatterPlot** with the **fit** option:

```
ScatterPlot(male, female, fit=[a*x+b,x] );
```

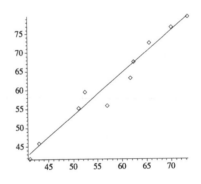

We can see how far each female life span is from its predicted span by asking for the **residuals**.

```
Fit(a*x+b, male, female,x, output=residuals);
```
[1.95239103558938432, 0.04735907620273622722, −1.09077780992713702, −3.04509298471736578, 0.864406712786381926, 1.02071967803876729, 0.464711839440238139, 0.972564905173992522, −5.03949680406133815, 3.85321435147434110]

Under this model, say, if a certain country's male life expectancy is 65 years, then the female life expectancy is estimated to be:

```
eval( line1, x=65 );
```
$$70.18318383$$

Testing Correlation

You use the **Correlation** command to find how well the best fitting line matches the data. Using our previous example:

```
Correlation( male, female );
```
$$0.9794660676$$

This value is very close to 1. Thus, a strong correlation exists between the life expectancy of males and females in the countries.

Fitting with Other Functions

The **Fit** command and option can also work for other functions. Given functions $f_1(x)$, $f_2(x)$, ..., $f_n(x)$, if you want to fit a function of the form

$$a_1 f_1(x) + a_2 f_2(x) + \ldots + a_n f_n(x)$$

that best matches a given set of data, you will use one of these formats:

Fit ($a_1 f_1(x) + a_2 f_2(x) + \ldots + a_n f_n(x)$, *x-data*, *y-data*, **x**);

For example, using the previous data:

```
Fit( a+b*x+c*x^2, male, female, x );   #best quadratic
```
$$5.74472328015188350 + 0.802963923918145639\ x$$
$$+ 0.00287185630353596188\ x^2$$

```
Fit( a+b*x+c*x^2+d*x^3, male, female, x);  #best cubic
```
$$-122.129348470905157 + 7.81009910122232842\ x$$
$$- 0.122245426757801548\ x^2 + 0.000729508526684393388\ x^3$$

```
Fit( a+b*exp(2*x)+c*exp(x), male, female, x );
```
```
Warning, model is not of full rank
```
$$1.71360581920718381\ 10^{-62}\ e^{2x} + 2.60283618819582845\ 10^{-94}\ e^x$$

The last answer suggests that it is not a good idea to use exponential functions to fit the given data in this case. Indeed, the coefficients of the exponential terms are virtually zero.

Using Weights in the Least Squares Method

A drawback with the least squares method of curve fitting is that it is extremely sensitive to a single bad data point. For fitting linear equations, one alternative is to weight the data.

■ **Example.** Consider the following linear fittings:

```
Fit( a*x+b, [20, 30, 40, 50], [39,  1, 79, 101], x );
```
$$2.63999999999999924\ x - 37.3999999999999986$$

```
Fit( a*x+b, [20, 30, 40, 50], [39, 61, 79, 101], x );
```
$$2.03999999999999915\ x - 1.39999999999997193$$

Notice how different these results are. In the first version, the second y-value was copied wrong. One bad point has affected the result dramatically.

One remedy is to give less weight to points considered to be suspect. The weights must be positive integers and the list of weights should be the same length as the list of points. In the following command, the "good points" are each weighted with 20, and the "bad point" is weighted with 1 only.

$$\text{Fit(} \texttt{a*x+b, [20, 30, 40, 50], [39, 1, 79, 101], x,}$$
$$\texttt{weights = [20, 1, 20, 20]);}$$

$$2.09790209790209747x - 4.87412587412584219$$

Adding one bad point with less weight than the others gives a line that resembles the line obtained without the point.

Interpolation

The Polynomial Interpolation Command

Given n points (x_1, y_1), (x_2, y_2), ..., (x_n, y_n), a second way of fitting a curve to the data is to find a polynomial of degree $n - 1$ that perfectly matches all these values. This can be done in Maple using the command **PolynomialInterpolation** from the **CurveFitting** package in the syntax:

```
with(CurveFitting):
PolynomialInterpolation( x-data, y-data, x);
PolynomialInterpolation( list of points, x);
```

For example, to find a polynomial $g(x)$ that passes through all of the following points, use:

```
points:= [ [0,0], [1,16], [2,-10], [3,28],
           [4,-30], [5,-12], [6,-12]]:
```

```
with(CurveFitting):
g := PolynomialInterpolation(points, x);
```

$$-\frac{299}{180}x^6 + \frac{299}{10}x^5 - \frac{1819}{9}x^4 + \frac{1897}{3}x^3 - \frac{162041}{180}x^2 + \frac{13733}{30}x$$

You can see that the graph of $g(x)$ passes through all the given points with:

```
with(plots):
pict1 := pointplot( points, symbol=solidcircle,
                    symbolsize = 15 ):
pict2 := plot( g, x=0..6 ):
display( [pict1, pict2] );
```

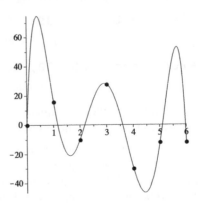

The Spline Command

A third approach for fitting a curve to n points (x_1, y_1), (x_2, y_2), ..., (x_n, y_n), is to look for a curve that is defined "piecewise" by polynomials of a given degree d, with the pieces agreeing up to the first $d-1$ derivatives at the connection points. This is done with the **Spline** command.

```
with( CurveFitting ):
Spline( list of points, x, degree = d );
```

We can compare the example of **Spline** and **PolynomialInterpolation** on the set of points below.

```
with( CurveFitting ):
data:= [[0,0],[1,10],[2,4],[3,10],[4,1],[5,10]]:
c3 := Spline( data, x, degree = 3 ):
c4 := PolynomialInterpolation( data, x ):
pict1 := plot( [c3,c4], x=0..6, y=-4..14,
                legend=["spline","poly"] ):

with(plots):
pict2 := pointplot( data, symbol=solidcircle,
                      legend="data points"):
display( [pict1, pict2] );
```

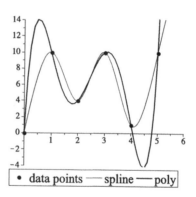

More Examples

Fitting vs. Interpolation

You may wonder about the difference between the methods of least squares fitting and interpolation. A least squares fit allows us to find a function of a specified form that best *approximates* the given set of data. On the other hand, interpolation will find a function that *matches the data exactly*. A least squares fit is normally used when we don't believe the independent variable entirely predicts the dependent variable.

■ **Example.** Consider the UNESCO data of the first example of this chapter. If we had used interpolation rather than finding a least squares fit for this data, we would have found the polynomial:

```
Digits := 12:        # We need to increase the number of digits used in
                     # the floating point computation, because the answer
                     # behaves so wildly, as you will see.
```

```
with(CurveFitting):
PolynomialInterpolation(male,female,x);
```

$$-3.97572075225 \cdot 10^{-8} x^9 + 0.0000203073369109 \, x^8 - 0.00458829289279 \, x^7$$
$$+0.601845366134 \, x^6 + 50.5044284571 \, x^5 + 2811.63405657 \, x^4 - 1.03837722472 \cdot 10^5 \, x^3$$
$$+2453065.00241 \, x^2 - 3.36364582455 \cdot 10^7 \, x + 2.03962232163 \cdot 10^8$$

Let's compare graphically the results obtained by the two methods:

```
with(Statistics):
pict1:=ScatterPlot(male, female, fit = [a*x+b, x]):

pict2 := plot(PolynomialInterpolation(male,female,x),
              x=40..75, y = 30..80, color = grey ):

with(plots);
display( [pict1, pict2] );
```

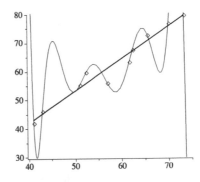

You can see in this case that the line obtained by least squares fitting is likely to be better for prediction.

A Temperature Model

■ **Example.** The following data shows the average monthly high temperatures in degrees Fahrenheit in Boston, Massachusetts for 15 months starting in January.

```
bostonHigh := [36.4, 37.7, 45.0, 56.6, 67.0,
               76.6, 81.8, 79.8, 72.3, 62.5,
               51.6, 40.3, 36.4, 37.7, 45.0]:
```

We expect that the average monthly high temperature is sinusoidal, and that it can be described nicely by trigonometric functions in the form

Temperature $= a + b\cos(kx) + c\sin(kx)$, where x is the month.

The temperature's period is 12 months, so we should certainly choose $k = \dfrac{2\pi}{12} = \dfrac{\pi}{6}$.

Thus, we need to find a function of the form

$$a + b\cos(\frac{\pi}{6}x) + c\sin(\frac{\pi}{6}x)$$

that best fits the temperature data:

```
bestfit := Fit(a+b*cos((1/6)*Pi*x)+c*sin((1/6)*Pi*x),
               [seq(i, i = 1 .. 15)], bostonHigh, x):
pict1 := plot( bestfit, x = 0 .. 16 ):
pict2 := ScatterPlot( [seq(i, i = 1 .. 15) ],
                      bostonHigh, symbol = cross,
                      symbolsize=20):
with(plots):
display([pict1, pict2]);
```

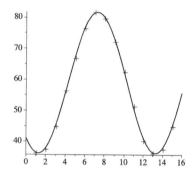

The temperature data quite nicely matches the function we've found using **Fit**. Also, we notice from the picture that the highest point of the graph is around $x = 7.3$, which suggests that July 9 will be the hottest day of the year.

Useful Tips

💡 💡   Fitting data to a curve can also be done with the **LeastSquares** command from the **CurveFitting** package.

💡   If you are planning to do a lot of statistics, we recommend that you also read Chapters 24 and 25, so you can learn more about lists, random numbers, and simulation.

Troubleshooting Q & A

Question...   I used the **Fit** command, but got error messages such as "`... 3rd argument must have type name,`" or "`... data points not in recognizable format.`" What went wrong?

Answer...   You probably have a problem with brackets. Either you forgot to put brackets around a list of points (giving too many arguments), or you put brackets around your lists of *x-data*, *y-data* (giving too few arguments).

Question...   I used the **Fit** command, but got error messages such as "...expects its 4th argument, v, to have type name, ..." What went wrong?

Answer...   You probably have already assigned the variable you used in fit. Use **unassign**.

Question...   I used the **ScatterPlot** command with the **fit** option to plot a data set with a curve, but got an obviously wrong flat line for the curve. What went wrong?

Answer...   You probably have already assigned the variable you used in fit. Use **unassign**.

Question...   I used the **Fit** command, but got error messages such as "...number of values in first and second arguments must match." What went wrong?

Answer...   Check that your lists of *x*-data and *y*-data have the same number of entries.

Question...   When I tried to graph the points of a data set along with the polynomial obtained from using **PolynomialInterpolation**, the curve did not go through the points. What went wrong?

Answer...   Inaccuracies produced by **PolynomialInterpolation** are often due to roundoff error in the calculation of the coefficients of the polynomial. You may need to increase the number of digits used in the floating point computation. For example, try: **Digits := 12;**

You can also use the **form=Lagrange** option.

Question...   Can you explain the difference between using the **LeastSquares** and **PolynomialInterpolation** commands?

Answer...   **LeastSquares** finds a function of a specified form that best *approximates* the given set of data. On the other hand, **PolynomialInterpolation** will find a function that *matches the data exactly*, although it may behave "wildly" in between the data points. See the first example of the "More Examples" section.

Question...   Why did I get an error message about "independent values must be in strictly ascending order" when I tried to use the **Spline** command?

Answer...   Since splines are piecewise defined between consecutive points, the data points must be sorted in terms of *x* before using the **Spline** command.

CHAPTER 24

Animation

An animation is a sequence of images displayed in succession. The illusion of a continuous motion can be obtained if consecutive images do not change too much. In this chapter you will learn two different ways to create an animation in Maple.

Getting Started

The animate Command

The **animate** command is the basic command for creating both 2-D and 3-D animations. This command is part of the **plots** package. Its general usage is:

```
with(plots);
animate( plot command, [plot arguments], t = t₀.. t₁ );
```

This will be explained in more detail shortly. For now, consider this example:

```
with( plots ):
animate( plot, [sin(t*x), x=0..2*Pi], t=1..7 );
```

Now, click on the picture. The context bar (near the top of the Maple window) will show a control strip with buttons similar to those on a DVD player. Click the "play" icon ▷ on the control strip to see the animation.

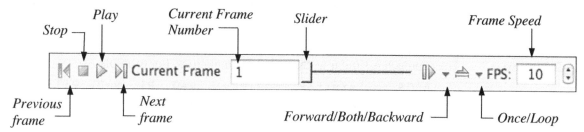

The animation that Maple constructs has, by default, 25 frames. When the animation first appears, the first frame is displayed. Notice that the value of the parameter, t, is reported in the title of each frame of the animation.

Constructing Animations with animate

The animation in the example above provides a nice visual illustration of the fact that, as a function of x, the graphs of $y = \sin(tx)$ all have amplitude 1 and periods that decrease as t increases. Before writing the **animate** command it is helpful to construct separate plots for several different values of the parameter t. For example:

```
plot( sin(x), x=0..2*Pi );        #Figure 1
plot( sin(2*x), x=0..2*Pi );      #Figure 2
plot( sin(3*x), x=0..2*Pi );      #Figure 3
```

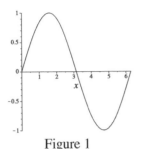
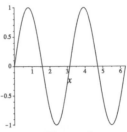
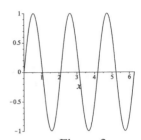

Figure 1   Figure 2   Figure 3

From here it is straightforward to come up with the appropriate format for the **animate** command. The parameter is **t**, and let **t** vary from 1 to 7. The *plot command* is **plot** and the *plot arguments* are **sin(t*x), x=0..2*Pi**. Notice how the only change from the previous **plot** statements is the appearance of the parameter **t**. This leads us to:

```
animate( plot, [sin(t*x), x=0..2*Pi], t=1..7 );
```

The next animation has frames that contain the graphs of six functions on the interval $-1 \leq x \leq 1$. The **animate** command works the same way as the **plot** command. The following demonstrate some variations of the **animate** command.

```
fnList := [t*x, -t*x, t*x^2, -t*x^2, t*x^3, -t*x^3]:
animate( plot, [fnList, x=-1..1], t=-1..2 );
```

In the next animation, each frame consists of two parametric circles with different (fixed) centers and varying radii.

```
c1 := [t*cos(x), t*sin(x), x=0..2*Pi]:
c2 := [(1-t)*cos(x)+2, (1-t)*sin(x), x=0..2*Pi]:
animate(plot, [[c1,c2], scaling=constrained],t=0..1);
```

Animating a 3-D Picture

The **animate** command can as well be used to create 3-D animations in Maple. The approach is similar to creating 2-D animations. For example,

```
ARGS := [sin(t*x)+cos(t*y), x=0..Pi, y=0..Pi]:
animate( plot3d, ARGS, t=1..5, axes=boxed );
```

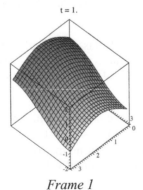
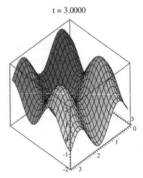

Frame 1   *Frame 13*

Two frames (frames 1 and 13) of the animation are shown. If you watch the animation, you will see "waves" move through the surface.

Advanced Examples

The display Command

The examples we have looked at so far have been simple animations in which each frame could be plotted with a single **plot** command. When we want to make more complicated animations, we will often make use of the **display** command to combine various graphics into a frame.

Strategy

The key to writing sophisticated animations is to have a firm grasp of what each frame looks like, as well as how the frames vary according to a parameter. Following this concept, you can more easily create a complicated animation.

In cases where it is not possible to create each frame with a single plot command, we'll define **oneFrame**, a "frame function" of t, that will generate the graphic elements of just one frame. Because each frame itself will be a combination of several simple graphics, we frequently use the **display** command within the definition of **oneFrame**:

```
oneFrame := t->display([ graphics command(s) depending on t ]);
```

Then we will use the **animate** command to create an animation built from a sequence of frames produced by **oneFrame**. If there is a common background found in every frame of the animation, this plot should be prepared separately and passed to **animate** as an optional **background** argument. A common form of the **animate** command is:

```
with(plots):
animate( oneFrame, [t], t= t₀.. t₁, background= BACK );
```

Moving a Point Along a Curve

■ **Example.** Let's show you how to write the animation of a point moving along the sine graph, $y = \sin(x)$.

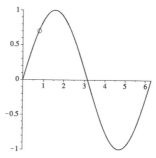

The coordinate of the point at any time t can be thought of as $P(t) = (t, \sin(t))$ for $0 \le t \le 2\pi$. In each frame of the animation, we will plot the position of the point at time t against a background of the sine curve. The sine curve can be sketched with a **plot** command

```
with(plots):
BACK := plot( sin(x), x=0..2*Pi ):
```

The point, at time t, can be drawn using the **pointplot** command. We can define

```
oneFrame := t -> pointplot([t,sin(t)], color=blue,
                symbol = circle, symbolsize = 18);
```

To animate over the interval $0 \le t \le 2\pi$, we use:

```
animate(oneFrame,[t],t=0..2*Pi,background = BACK);
```

Rolling a Ball Along the Ground

■ **Example.** A circle of radius one "rests" on top of the x-axis at the origin, as pictured below. A point P, which is marked at the "top" of the circle, is currently touching the y-axis at the point $(0, 2)$.

As the circle begins to "roll" to the right, the point P will rotate downward and eventually hit the x-axis after the circle has rolled a distance of π (because P was originally halfway around a circle of circumference 2π). Thereafter, P will rotate upward again after the circle "rolls over" it, reaching a high point again when the circle has rolled a distance of 2π.

The following suggests the motion of P, where a dot shows the location of P:

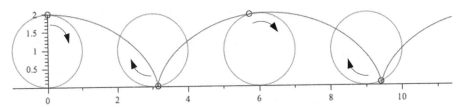

In fact, the path followed by P as the circle rolls is known to be a cycloid, which can be given parametrically by $(u + \sin(u), 1 + \cos(u))$, for $0 \le u$.

We will let the parameter t in this animation be the distance the ball has rolled. Each frame of the animation should have three elements:

- A fixed portion of the cycloid to act as a background. It can be given by $(u + \sin(u), 1 + \cos(u))$, for $0 \le u \le 4\pi$.

- The new position of the unit circle in this frame. After the circle moves a distance t, its center is at $(t, 1)$ and its parametric equation is $(t + \cos(u), 1 + \sin(u))$ for $0 \le u \le 2\pi$.

- The corresponding position of the point P. After the ball has traveled a distance t, P will have rotated a distance of t clockwise around the circle. P's location relative to the center of the circle is thus given by $(\sin(t), \cos(t))$. Its position in the frame will then be $(t + \sin(t), 1 + \cos(t))$.

After some experimentation with **view** to adjust the range of display for the x- and y- coordinates, our frame function is as follows.

The cycloid is plotted once, to be passed to **animate** as an optional argument.

```
with(plots):
BACK := plot( [u+sin(u), 1+cos(u), u=0..4*Pi],
              color=blue, scaling=constrained ):
```

The circle and the point P change with the parameter t. A separate procedure, **oneFrame**, uses **display** to combine these two objects into a single plot.

```
oneFrame := t -> display( [
    plot( [t+sin(u), 1+cos(u), u=0..2*Pi] ),
    pointplot( [t+sin(t), 1+cos(t)], symbol=circle)
                    ] ):
```

The completed animation, which rolls P over two arcs of the cycloid, is given with:

```
animate( oneFrame, [t], t=0..4*Pi, background=BACK );
```

**The
Oscillating
Spring**

■ **Example**. A metal plate is attached to a spring of height h_0. (The picture to the right shows such a spring at a height of 4.2 units above the xy-plane.) The plate is then displaced upward a distance of c units and released. The plate (and spring) will begin a vertical, damped oscillation.

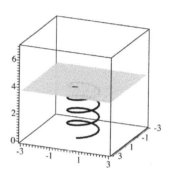

If we ignore the effect of gravity and the weight of the plate, then the height of the plate at any time t after release will be

$$z = h_0 + c\,e^{-bt}\cos(kt),$$

where b and k are constants that depend on the spring.

We'll model the motion of the plate using animation in Maple. The graphic elements of each frame are as follows.

- The plate, at height $z = h_0 + c\,e^{-bt}\cos(kt)$ can be drawn with **plot3d**. We'll sketch it only over the rectangle $-1 \le x \le 1$, $-1 \le y \le 1$ and use **transparency =0.5** to ensure the spring is visible through the plate.

- We'll think of the "spring" as the helix $(\cos(u), \sin(u), \dfrac{u}{8\pi}(h_0 + ce^{-bt}\cos(kt)))$, for $0 \le u \le 8\pi$, and sketch it using **spacecurve**.

The following "frame function" accepts the parameter t, (although it depends on the other global parameters h_0, c, b, and k) and returns a plot showing the plate and spring.

```
with(plots):
oneFrame := t -> display(
    [spacecurve( [cos(u),sin(u),
        u* (h0+c*exp(-b*t)*cos(k*t))/8/Pi],
        u=0..8*Pi, numpoints = 80 ),
    plot3d(h0+c*exp(-b*t)*cos(k*t),
        x=-3..3, y=-3..3, style=surface,
        color=cyan, transparency=0.8 )],
axes=boxed );
```

The motion of this plate and spring, with $h_0 = 5$, $c = 2$, $b =0.05$, and $k = 1$ can be seen with:

```
h0 :=5; c:=2; b:=0.05; k:=1;
animate( oneFrame, [t], t=0..25);
```

Try repeating the animation for different numerical values of the constants and see how each one affects the movement of the plate. For example, increasing the value of b will cause the oscillation to die out more quickly.

Useful Tips

 The **plot** command has many options. Some, like **style**, apply to the object being graphed. Others, like **titlefont**, apply to background settings. However, the **display** command cannot reconcile conflicting background settings from the various plot structures that make up an animation. It simply chooses some and ignores others. It's usually most efficient to specify these background settings once, as optional arguments in the **animate** command.

 The **animate** command accepts most of the same options as **plot** and **plot3d**. We have already seen the **background** argument for adding a static background to all frames of the animation. In addition, use **frames=**n to specify the number of frames in the animation (the default is 25 frames). For more options, and additional details, please search online help for **plots,animate** commands.

You can also create animations using a list of plots. The general way to do this is with the display command, and the optional argument **insequence=true** . For example, let's create separate plots of one period of the sine and cosine functions

```
S := plot( sin(x), x=-Pi..Pi ):
C := plot( cos(x), x=-Pi..Pi ):
```

The two graphs can be displayed in a single plot using **display** as follows:

```
display( [S,C] );
```

Simply adding **insequence=true** converts this into a two-frame animation:

```
display( [S,C], insequence=true);
```

Troubleshooting Q & A

Question... How do I tell Maple I want the animation to contain the frames for specific values of the parameter?

Answer... When the parameter values are given as a range, **t=**t_0 **..** t_1, Maple uses a uniform sampling of parameter values from this range. If, however, the second argument to animate is of the form **t=L** where **L** is a list of values for the parameter, then Maple uses these specific values of **t** to form the animation. For example,

```
animate(plot, [x^t,x=0..1], t=[1/4,1/3,1/2,1,2,3,4]);
```

CHAPTER 25

More About Lists

We have been using lists throughout this text. (See Chapter 6.) In this chapter we will show you more commands that are useful in working with lists. Many of them will be helpful in statistics or Maple programming.

List Basics

Lists and their Elements

A list is an ordered sequence of Maple objects enclosed in square brackets. The entries of a list can be any valid Maple object. For example, consider the list,

```
myList := [ "red", 2, 3, 3.153, x^2=5, Pi, sin, Pi,
            "history", `a Maple name`];
```

$$[\text{"red"}, 2, 3, 3.153, x^2 = 5, \pi, \sin, \pi, \text{"history"}, a \ Maple \ name]$$

Elements of **myList** include strings, integers, real numbers, an equation, and a function. Element can be referenced by **myList[1]**, **myList[2]**, and so on.

```
myList[5];
```

$$x^2 = 5$$

The syntax **myList[-n]** denotes the element of **myList** that's *n* entries *from the end* of the list.

```
myList[-3];
```

$$\pi$$

Use **myList[*a* .. *b*]** to specify a range of entries of **myList**.

```
myList[3..6];
```

$$[3, 3.153, x^2 = 5, \pi]$$

Note that the result of **myList[*a..b*]** is a list. If you want to see the elements of a list as a sequence of expressions, you can add **[** *empty square bracket* **]** after the list:

```
myList[3..6][];
```

$$3, 3.153, x^2 = 5, \pi$$

If an element of a list is itself a list, we can retrieve its entries in the same manner, using [*square bracket*] syntax. For example,

```
newList := [[1, 5, 7], [-6, 5, 7,7], 4, 6, 8]:
newList[2];
```

$$[-6, 5, 7, 7]$$

```
newList[2][1..3];        # The first three elements of newList[2]
```

$$[-6, 5, 7]$$

Modifying the Elements of a List

To change an entry in a list, you can use a normal assignment statement. Recall the list that we used earlier.

```
myList;
```

$$[\text{"red"}, 2, 3, 3.153, x^2 = 5, \pi, \sin, \pi, \text{"history"}, a\ Maple\ name]$$

We want to replace the first element of this list, say, with [26, 27]:

```
myList[1] := [26,27];
```

$$[26, 27]$$

Now you can see that the first element of **myList** has been replaced.

```
myList;
```

$$[[26, 27], 2, 3, 3.153, x^2 = 5, \pi, \sin, \pi, \text{"history"}, a\ Maple\ name]$$

We can also modify a list directly by redefining its elements, for example:

```
myList := [ myList[1..3][], true, false,
            myList[5..-1][], 100, 101 ];
```

$$[[26, 27], 2, 3, true, false, x^2 = 5, \pi, \sin, \pi, \text{"history"}, a\ Maple\ name, 100, 101]$$

The new **myList** consists of the first three elements of the original list, then the new elements **true** and **false**, followed by the fifth through last elements of the original list, and then the numbers 100 and 101.

■ **Example.** It might be of interest to construct a list of the first 1000 prime integers. We find these with the **ithprime** command

```
LotsOfPrimes := [ seq(ithprime(i), i=1..1000 ) ]:
```

We can see the first three and last three elements of this list with:

```
[ LotsOfPrimes[1..3][], LotsOfPrimes[-3..-1][] ];
```

$$[2, 3, 5, 7901, 7907, 7919]$$

You were probably already familiar with the first few primes, but not many people realize that 7919 is the 1000th prime.

Useful List Commands

The map Command

To evaluate a function at each element of a list, you use the **map** command. (The function can be Maple's built-in function or your own defined function.) The form of this command is:

map(*function name* **,** *the list* **);**

Let's look at an easy example, factoring polynomials of the form $x^n - 1$:

list1 := [seq(x^n-1, n=1..4)];

$$[x-1,\ x^2-1,\ x^3-1,\ x^4-1]$$

map(factor, list1);

$$\left[x-1, (x-1)(x+1), (x-1)\left(x^2+x+1\right), (x-1)(x+1)\left(x^2+1\right)\right]$$

Note: The first argument of the **map** command is only the name of the function or command you wish to evaluate at each element of a list. It should have no parentheses or arguments itself.

Now, we'll try a more interesting example:

grades := [[`John`, 75, 62], [`David`, 62, 81],
**   [`Mary`, 75, 91], [`Jane`, 31, 50],**
**   [`Steve`,21, 31]]:**

This shows a list of five students and their scores on two exams. The **avg** function, defined next, calculates the average score of each student and lists it with the name:

avg := x -> [x[1], (x[2]+x[3])*0.5]:

(Note that **x[1]** is the student's name, while **x[2]** and **x[3]** are the exam scores.)

grades2 := map(avg, grades);

$$[[John, 68.5], [David, 71.5], [Mary, 83.0], [Jane, 40.5], [Steve, 26.0]]$$

You can also use **map** to evaluate a multi-variable function with the first argument coming from the elements of a list.

map(*function/command name,* *the list,* *argument 2,* *argument 3,* *etc.***);**

For example, we define a list of x^n with

Flist := [seq(x^n, n=1..8)];

$$\left[x, x^2, x^3, x^4, x^5, x^6, x^7, x^8\right]$$

We can then compute the 2nd derivatives **diff(x,x,x), diff(x^2,x,x),** **diff(x^3, x, x)** etc. with:

map(diff, Flist, x, x);

$$\left[0, 2, 6x, 12x^2, 20x^3, 30x^4, 42x^5, 56x^6\right]$$

The map2 Command

Similarly, there is a **map2** command where a list is used for the second argument of a multi-variable function.

> **map2(** *function name* **,** *argument*₁ **,** *the list* **,** *argument*₃ **,** *etc.* **);**

In the next example we show how to solve (**isolate**) for each of the four variables that appears in the equation for a straight line ($y=mx+b$).

> **map2(isolate, y=m*x+b, [x,y,m,b]);**

$$\left[x = -\frac{-y+b}{m}, \ y = mx+b, \ m = -\frac{-y+b}{x}, \ b = y-mx \right]$$

The **isolate** command is similar to **solve**, except that it always returns the answer in the form of an equation.

The **map** and **map2** commands provide an easy way to implement "data transformation" or "data massaging" if you work with statistical data. See the US population model in the More Examples section.

The zip Command

The **map** and **map2** commands allow only one of the arguments of a multi-variable function to come from a list. The **zip** command can be used if both of the arguments of a two-variable function are lists.

> **zip(** *function name* **,** *first list* **,** *second list* **);**

For example,

```
Prod := (x, y) -> x*y:
Lista := [1,2,3,4,5]:
Listb := [6,7,8,9,10]:

zip( Prod, Lista, Listb );
```
$$[6, 14, 24, 36, 50]$$

We can also use the **zip** command to combine the lists **Lista** and **Listb** above into a list of ordered pairs:

> **zip((x,y)->[x,y], Lista, Listb);**

$$[[1, 6], [2, 7], [3, 8], [4, 9], [5, 10]]$$

Sorting Lists

The **sort** command will arrange the elements of a list in increasing order. It has the syntax:

> **sort(** *list* **);**

For example,

```
Nlist := [81, 70, 97, 63, 76, 38, 85, 68, 21]:
sort( Nlist );
```
$$[21, 38, 63, 68, 70, 76, 81, 85, 97]$$

If your list does not have elements that are numbers, or you want to apply a different ordering to the elements of a list, you can specify any Boolean function of two variables as the sorting criterion. It has the syntax:

> **sort(** *list* **,** *Boolean function* **);**

For example, to sort the **Nlist** above in descending order it is necessary to define a Boolean function that accepts two arguments and returns *true* if the first is larger than the second, and *false* otherwise. Here, this function is called **descending**.

```
descending := (a,b) -> evalb( a>b ):
sort( Nlist, descending );
```

[97, 85, 81, 76, 70, 68, 63, 38, 21]

This flexibility lets us sort lists whose elements are lists themselves. Consider the following example that sorts our list of students according to the average of their two exam scores:

```
grades;          #We defined the list earlier.
```

[[*John*, 75, 62], [*David*, 62, 81], [*Mary*, 75, 91], [*Jane*, 31, 50], [*Steve*, 21, 31]]

```
compareE2 := (a,b) -> evalb( a[3] > b[3] ):
```

(Note that in this function, **a** and **b** each refer to a list **[** *name, exam1, exam2* **]**; thus **a[3]** and **b[3]** are the third elements, *exam2*, of two different students.)

```
sort( grades, compareE2);
```

[[*Mary*, 75, 91], [*David*, 62, 81], [*John*, 75, 62], [*Jane*, 31, 50], [*Steve*, 21, 31]]

The select and remove Commands

The **select** command selects all elements of a list that meet a specified Boolean condition. In the same way, the **remove** command removes all elements of a list that meet a Boolean condition. These commands are typically used with the following syntax:

```
select( Boolean condition, list );
remove( Boolean condition, list );
```

For example, here is one way to construct lists of prime and composite integers:

```
myIntegers := [ $1..20 ]:      #a list of integers from 1 to 20
P := select( isprime, myIntegers );
```

[2, 3, 5, 7, 11, 13, 17, 19]

```
C := remove( isprime, myIntegers );
```

[1, 4, 6, 8, 9, 10, 12, 14, 15, 16, 18, 20]

In situations like this when you really want to separate a list into two sublists, the **selectremove** command is particularly convenient. These same lists of prime and composite integers can be formed in one command as follows:

```
P2,C2 := selectremove( isprime, myIntegers ):
P2;
```

[2, 3, 5, 7, 11, 13, 17, 19]

```
C2;
```

[1, 4, 6, 8, 9, 10, 12, 14, 15, 16, 18, 20]

> **Note:** The left-hand side of the assignment consists of two names separated by a comma. The two lists produced by **selectremove** are separately assigned to these two names.

Counting Elements

The **Tally** command in the **Statistics** package shows how many times each item appears in a list. It has the syntax:

>   Statistics[Tally](*your list*);

For example,

>   list1 := [3, 3, 5, 6, 5, 9, 10, 5, 5, 1, 5, 1, 5];
>
> $$[3, 3, 5, 6, 5, 9, 10, 5, 5, 1, 5, 1, 5]$$
>
>   Statistics[Tally](list1);
>
> $$[1 = 2, 3 = 2, 5 = 6, 6 = 1, 10 = 1, 9 = 1]$$

This means that 1 and 3 each appear twice, 5 appears six times, and 6, 10, and 9 appear only once each.

Other List Commands

Many other commands for lists will be useful, especially for programming in Maple. We list a few of them here. Some of the commands are in the **ListTools** package.

Maple Command	*Explanation*

`max( [ 2, 5, -3, 1.2 , 6.01, 7.5][] );`
7.5
`min( [ 2, 5, -3, 1.2 , 6.01, 7.5][] );`
−3
	max and **min** are used to find the largest and smallest elements of a comma-separated sequence of numbers.

`nops( [ 2, 5, -3, 1.2 , 6.01, 7.5] );` 6	**nops** tells you the number of elements in a list or set.

`with( ListTools ):`
`Reverse( [1, 3, 5, 7, 9] );`
[9, 7, 5, 3,1]
	Reverse will reverse the order of the elements in a list. The first element becomes the last and vice versa.

`with(ListTools):`
`Rotate( [1, 2, 3, 4, 5], 2 );`
[3, 4, 5, 1, 2]
	Rotate will rotate the elements of the list to the left in a circular way.

`with(ListTools):`
`Transpose([[1,3,5,7], [2,4,6,8]]);`
[[1, 2], [3, 4], [5, 6], [7, 8]]
	Transpose operates on a list of lists as if it were transposing a matrix. This is most useful when changing a pair of lists into a list of ordered pairs.

`with(ListTools):`
`Flatten([[1,3,5,7], [2,4,6,8]]);`
[1, 3, 5, 7, 2, 4, 6, 8]
	Flatten combines the entries of several lists into a single list.

More Examples

Baseball Standings

■ **Example.** The following list represents the final standings of the American League Eastern Division Baseball teams for the 2008 regular season in the form [*team, wins, losses*]:

>   baseball := [[`Bal`, 68, 93], [`Bos`, 95, 67],
>   [`NY`, 89, 73], [`TB`, 97, 65], [`Tor`, 86, 76]]:

When you look at the sports page of a newspaper, you'll see that the team standings include the team's winning percentage (PCT) and the number of games behind (GB). Also, teams with a higher winning percentage will be listed first. Let's show how to do this in Maple:

- We first add the winning percentage to each element (team). The **addWP** function appends the winning percentage, to 3 decimal places, to the existing data for each team.

```
addWP := x -> [ x[], evalf[3](x[2]/(x[2]+x[3])) ]:
bb2 := map( addWP, baseball );
```

$$[[Bal, 68, 93, 0.422], [Bos, 95, 67, 0.586], [NY, 89, 73, 0.549],$$
$$[TB, 97, 65, 0.599], [Tor, 86, 76, 0.531]]$$

- Next we **sort** the list in descending order of the winning percentage.

```
compareWP := (a,b) -> evalb( a[4]>b[4] ):
bb3 := sort( bb2, compareWP );
```

$$[[TB, 97, 65, 0.599], [Bos, 95, 67, 0.586], [NY, 89, 73, 0.549],$$
$$[Tor, 86, 76, 0.531], [Bal, 68, 93, 0.422]]$$

- The "games behind" is the number of games a team would have to win (against the first-place team) in order to move into a tie with the first-place team.

```
addGB := x -> [ x[],
              (bb3[1][2]-x[2]+x[3]-bb3[1][3])*.5 ]:
bb4 := map( addGB, bb3 );
```

$$[[TB, 97, 65, 0.599, 0.], [Bos, 95, 67, 0.586, 2.0], [NY, 89, 73, 0.549, 8.0],$$
$$[Tor, 86, 76, 0.531, 11.0], [Bal, 68, 93, 0.422, 28.5]]$$

- Finally, to display the standings with each team line by line, just like you see it in a newspaper, we display this list as a matrix:

```
Matrix( bb4 );
```

$$\begin{bmatrix} TB & 97 & 65 & 0.599 & 0. \\ Bos & 95 & 67 & 0.586 & 2.0 \\ NY & 89 & 73 & 0.549 & 8.0 \\ Tor & 86 & 76 & 0.531 & 11.0 \\ Bal & 68 & 93 & 0.422 & 28.5 \end{bmatrix}$$

A U.S. Population Model

■ **Example.** The U.S. population (measured in millions) over the last two hundred years, between 1790 and 1990, is given in ten-year increments as follows:

```
year :=[ seq( 1790 + i*10, i=0..20 ) ];
```

[1790, 1800, 1810, 1820, 1830, 1840, 1850, 1860, 1870, 1880, 1890, 1900, 1910, 1920, 1930, 1940, 1950, 1960, 1970, 1980, 1990]

```
population := [ 3.929, 5.308,  7.240, 9.638, 12.861,
        17.063, 23.192, 31.443, 38.558, 50.189, 62.980,
        76.212, 92.228, 106.021, 123.203, 132.166,
        151.326, 179.323, 203.302, 226.542, 248.710]:
```

Theoretically, this population follows a "logistic model" and hence has the form:

$$population = \frac{288.5}{1 + e^{a+bt}} \text{ , where } t \text{ is the year.}$$

(The constant 288.5 represents a theoretical maximum sustainable population of 288.5 million, which can be predicted by the data.) We want to find the constants a and b so that the logistic model best fits the data.

We cannot use the **LeastSquares** command (discussed in Chapter 23) to find a and b directly. However, one can easily rewrite the expression

$$population = \frac{288.5}{1 + e^{a+bt}} \text{ as } \frac{288.5}{population} = 1 + e^{a+bt} \text{ , and thus we have:}$$

$$\ln\left(\frac{288.5}{population} - 1\right) = a + bt \text{ .}$$

This suggests that we can find a and b from the data $\ln\left(\frac{288.5}{population} - 1\right)$. Thus, we

must transform each of the given data pairs $[t, P]$ to $\left[t, \ln\left(\frac{288.5}{population} - 1\right)\right]$.

```
poptrans := map(p->ln(288.5/p-1),population);
```

[4.282597842, 3.976909996, 3.659658302, 3.365003418, 3.064892562, 2.766817692,
    2.437084024, 2.101121467, 1.869065290, 1.557780664, 1.275591631, 1.024424913,
    0.7552376804, 0.5429979104, 0.2939104542, 0.1679360362, –0.09818625675,
    –0.4962182298, –0.8697145978, –1.296473621, –1.832671937]

```
CurveFitting[LeastSquares]( year, poptrans, t );
```

$$55.32447502 - 0.0285529405t$$

This suggests that $a = 55.3245$ and $b = -0.0285529$. The logistic model is thus:

$$population = \frac{288.5}{1 + e^{55.3245 - 0.0285529t}} \text{ , where } t \text{ is the year.}$$

```
P := t -> 288.5/(1+exp(55.32447502-0.02855294054*t)):
plot1 := Statistics[ScatterPlot]( year, population ):
plot2 := plot( [P(t),288.5], t=1790..2020 ):
plots[display]( [ plot1, plot2 ] );
```

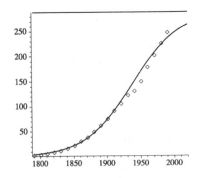

The logistic model fits the data very nicely, especially through about 1930.

Useful Tips

 Maple's internal representation of lists is different for lists with more than 100 elements. While subscript notation can be used to access elements of such long lists, we cannot use subscript notation to change values of an element of a long list. For example,

```
LL := [ $1..101 ]:
LL[3];
```

$$3$$

```
LL[3] := "a new value";
```

```
Error, assigning to a long list, please use Arrays
```

The **subsop** command provides an efficient way to change individual entries in a long list. For example, to change the third element of the list you would use:

```
LL2 := subsop(3="a new value",LL):
LL2[1..10];
```

$$[1, 2, \text{"a new value"}, 4, 5, 6, 7, 8, 9, 10]$$

The **subsop** command is also useful for removing an entry from a long list:

```
LL3 := subsop(3=NULL,LL):
LL3[1..10];
```

$$[1, 2, 4, 5, 6, 7, 8, 9, 10]$$

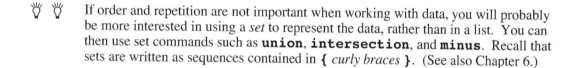 If order and repetition are not important when working with data, you will probably be more interested in using a *set* to represent the data, rather than in a list. You can then use set commands such as **union**, **intersection**, and **minus**. Recall that sets are written as sequences contained in **{** *curly braces* **}**. (See also Chapter 6.)

CHAPTER 26

Random Numbers and Simulation

Random Numbers

The rand Command

We can use Maple to generate sequences of random numbers. Such sequences can be used to simulate probabilistic situations.

The **rand** command generates a random integer between the numbers a and b.

```
rand( a .. b )();
```

If you need a random integer between 0 and $n-1$ (inclusive), you can also enter:

```
rand( n )();
```

For example, you want a random integer between 1 and 6:

```
rand( 1..6 )();
```
$$5$$

For a random integer between 0 and 9,

```
rand( 10 )();
```
$$6$$

The RandomTools Package

The **RandomTools** package provides another way to generate random objects. The basic syntax for obtaining a random integer between the numbers a and b is:

```
with( RandomTools ):
Generate( integer( range = a .. b ) );
```

For example, you want a random integer between 1 and 6:

```
with( RandomTools ):
Generate( integer( range=1..6 ) );
```
$$5$$

Whereas the **rand** command produces only integers, the **Generate** command can produce a wide variety of random objects. This will be illustrated in the examples in the rest of this chapter.

Pseudo-Randomness

Sequences of numbers returned by **rand** and **Generate** are said to be "pseudo-random." They are actually generated using an iteration formula. They eventually repeat but not soon enough so that you would notice.

The next graph plots 1000 numbers between 0 and 99 obtained from **rand**. It certainly "looks random."

```
with(plots):
datapoints := [seq([i, rand(100)()], i=1..1000)]:
pointplot( datapoints );
```

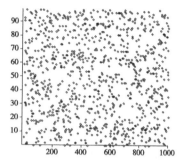

There seems to be no obvious pattern emerging, confirming a sense of randomness. However, by default, every time you start Maple, you will get the same sequence of random numbers. Thus

```
restart;
rand( 10^10 )();
```

$$581869302$$

always gives the same answer every time both commands are executed. The **randomize** command is used to shuffle the pseudo-random sequence.

```
restart;
randomize():          # uses the system clock to set the seed
rand( 10^10 )();      # returns a different random number
```

$$2492709555$$

■ **Example.** Two ways to generate 5000 random integers between 1 and 10 are:

```
data1 := [ seq(rand(1..10)(), i=1..5000) ] :
with( RandomTools ):
data2 := Generate(list( integer(range=1..10),5000 )):
```

If these 5000 values were truly random, we would expect each of the numbers 1, 2, . . . , 10 to show up about the same number of times. Let's check it:

```
Statistics[Tally]( data1 );
```

$$[1 = 487, 2 = 474, 3 = 511, 5 = 487, 4 = 519, 7 = 550, 6 = 522, 10 = 458, 8 = 478, 9 = 514]$$

```
Statistics[Tally]( data2 );
```

$$[1 = 504, 2 = 478, 3 = 449, 5 = 529, 4 = 507, 7 = 516, 6 = 551, 10 = 491, 8 = 494, 9 = 481]$$

Notice that, for each of the two lists of 5000 integers, each digit appeared almost equally often (close to 500 times). Consequently, we see that **Generate** and **rand** produce a uniform distribution of numbers between 1 and 10.

Examples in Simulation

To simulate a real-world experiment using the **rand** or **Generate** command requires a little programming. Here is the general approach:

- Write a procedure that generates the result of a single experiment. The procedure is written in the form:

  event := **proc ()**
      local *local variables separated by commas* **;**
      Maple commands separated by semicolons or colons **;**
  end proc;

  In many cases the procedure involves the **if** command, usually in the form:

  if *condition* **then** *result1* **else** *result2* **end if;**

  Maple will give *result1* if *condition* is *true*. If *condition* is *false*, *result2* is returned. (The **proc** and **if** commands will be discussed in detail in Chapter 27.)

- Then, we use the **seq** command to repeat the experiment *n* times:

  data := [seq(*one result of a random experiment*, **i = 1..***n***)];**

- After simulating the experiment, we'll analyze the data and summarize what we observed. This often involves using commands from the **Statistics** package (discussed in Chapter 22).

Flipping a Coin

■ **Example**. The flipping of a "fair" coin will give us either a head or a tail, with both results being equally probable. That is, approximately 50% of the time we'll get a head, and 50% of the time we'll get a tail. Since **rand(0..1)()** gives a random integer 0 or 1, we can simulate flipping a coin with:

```
flip := proc()
  local HT;
  HT:= rand(0..1);
  if HT() = 0  then  Head  else  Tail  end if;
end proc:
```

This guarantees that every time Maple evaluates **flip()**, you get a new random number, corresponding to a new flip of the coin.

flip(),flip();
               Head, Tail

Here is a list of eight coin flips:

[seq(flip(), i=1..8)];
               [Tail, Tail, Tail, Tail, Head, Head, Head, Tail]

This simulation is a little easier to implement with **Generate**. We setup the structure of the object we want to create (a head or a tail) using the **choose** command:

flip2 := choose({Head,Tail}):

Then, we can generate the outcome of a single event

```
Generate( flip2 );
```
                Head

Here is another list of eight coin flips:

```
Generate( list( flip2, 8 ) );
```
              [*Head, Tail, Tail, Head, Head, Head, Head, Tail*]

Notice how much easier, and more natural, it is to use **Generate** than **rand**.

Let's Roll a Die

■ **Example**. A standard die has six sides labeled 1, 2, 3, 4, 5, and 6. If the die is "fair," we expect each number to appear on the top face about one-sixth of the time. We can simulate rolling a die with:

```
roll := rand( 1..6 ):
roll();
```
                  1
```
roll();
```
                  6

Now, suppose that players *A* and *B* each roll a die. The one who gets the larger number wins the game. Here is one way to simulate this game:

```
game := proc()
   local player1, player2, roll;
   roll := rand(1..6);
   player1 := roll():
   player2 := roll():
   if player1 > player2 then "A wins"
      elif player1 = player2 then "Tie"
      else "B wins"
   end if;
end proc:
```

Each time you evaluate the command **game()**, you get a new result.

```
game();
```
              "B wins"
```
game();
```
              "Tie"

Let us repeat this game 1000 times and count how many times each player wins.

```
data := [seq( game(), i=1..1000 )]:
Statistics[Tally]( data );
```
          ["Tie" = 136, "A wins" = 439, "B wins" = 425]

Based on this data, it appears that this is a "fair game" and that player A had slightly better luck this time. The implementation of this experiment with **Generate** is very similar.

Poker Anyone?

■ **Example**. A standard poker hand consists of 5 cards dealt from a 52-card deck. The deck consists of 13 cards (A, 2, 3, 4, 5, 6, 7, 8, 9, 10, J, Q, K) in four suits (clubs, diamonds, hearts, and spades). We can pick a single card from the deck with:

```
card := choose( {A,2,3,4,5,6,7,8,9,10,J,Q,K} )
        * choose( {c,d,h,s} ):
Generate( card );
```

$$6\,h$$

This means the card selected was the six of hearts.

Here's how to get a 5-card poker hand:

```
Generate( list( card, 5 ) );
```

$$[A\,d,\,K\,h,\,2\,c,\,K\,d,\,9\,h]$$

A pair of kings and an ace! Are you playing this hand?

A Strange Behavior of Numerical Answers

■ **Example.** Look at the answer sections in your physics, chemistry, or economics textbooks. Do you notice that the first digit of each numerical answer is more likely to be 1, 2, 3, or 4 rather than 5, 6, 7, 8, or 9? This is because after each multiplication or division of two numbers, the first digit of the result is more likely to be 1, 2, 3, or 4 than the others. (Most scientific values arise from multiplications and divisions.)

This phenomenon is related to properties of the logarithm function. If you don't believe it, let's do an experiment to demonstrate it.

We will choose two integers randomly between 1 and 100, multiply them together, and record the first digit of the product. It is tricky to extract the first digit of the product. We first **convert** the product into a string, use **[1]** to extract the first character, and then use **parse** to convert this string to a digit.

```
game := proc()
  local p:
  p:= RandomTools[Generate]( integer(range=1..100)
                           * integer(range=1..100) );
  parse( convert( p, string )[1] );
end proc:
```

Now we repeat this experiment 5000 times, and tabulate the results.

```
data := [seq( game(), i=1..5000 )]:
Statistics[Tally]( data );
```

$$[1 = 1116, 2 = 712, 3 = 576, 5 = 419, 4 = 602, 7 = 427,$$
$$6 = 428, 8 = 354, 9 = 366]$$

In 5000 experiments, we can see that the first digit of the result being 1, 2, 3, or 4 happens 1116+712+576+602 = 3006 times. Are you convinced?

A Game of Risk™, Anyone?

■ **Example.** In the Game of Risk™, players attempt to control a map of the world by occupying countries with their armies. In each turn of the game, a player (the "attacker") may choose to engage another player (the "defender") who occupies an adjacent country in battle. If the attacker eliminates the armies of the defender, the attacker takes over the country.

To simulate a battle, the attacker is allowed to roll two dice, but the defender only one. If the attacker's larger die is higher than the defender's die, the attacker wins the battle. However, the defender wins if his die is higher than or ties the larger die of the attacker.

The attacker has the advantage of rolling more dice, but the defender wins ties. So, which player has the advantage?

To find out, we can simulate the action of the attacker by rolling two dice and selecting the largest.

```
roll := rand( 1..6 ):
attacker := () -> max( roll(), roll() ):
```

The action of the defender is easier to simulate:

```
defender := () -> roll():
```

A sample battle might look like this:

```
attacker();
```
$$3$$

```
defender();
```
$$2$$

In this case, the attacker wins.

The outcome of every battle is either "attacker wins" or "defender wins". Thus we can simulate a battle with:

```
battle := proc ()
  if defender() < attacker() then
    "Attacker wins"
  else
    "Defender wins"
  end if
end proc:
```

Now, we can conduct 1000 battles and summarize with:

```
data := [seq( battle(), i=1..1000 )]:
Statistics[Tally]( data );
```
$$["Attacker wins" = 582, "Defender wins" = 418]$$

The result clearly shows that the attacker has the advantage over time. (In fact, the theoretical probability of the attacker winning over the defender is $\left(\frac{5}{6}\right)^3 \approx 57.87\%$. Our simulation is quite representative at 582/1000 = 58.2%.)

Happy Birthday to YOU!!

■ **Example**. The "birthday problem" is a famous problem in probability. For example, in a room of 30 people, how likely is it that at least two of them have the same birthday? We'll suggest an answer using Maple.

It's sensible to assume that birthdays of 30 randomly assembled people are spread evenly over the year, so we can simulate by picking 30 birthdays at random from 365 days of the year (sorry, we don't do leap years in this experiment!)

```
room := n -> RandomTools[Generate](
              list( integer( range=1..365 ), n ) ):
```

```
room( 30 );
```
$$[313, 205, 147, 260, 253, 303, 31, 87, 297, 285, 283, 100, 298, 3, 348, 133,$$
$$297, 27, 315, 356, 236, 72, 144, 166, 273, 333, 268, 258, 7, 95]$$

The output above simulates a typical "room" of 30 people, represented by their birthdays. In this collection of dates, notice that the number 297 appears twice, indicating that two people of this group had the same birthday, on the 297th day (October 24).

We can count how many *distinct* birthdays there are in the room of 30 people with:

```
nops( {room( 30 )[]} );
```
$$29$$

(The command `{room()[]}` converts the list of birthdays into a set and hence *removes duplicates*. `nops( {room(30)[]} )` is then the number of distinct birthdays.)

When `nops( {room(30)[]} ) < 30` is *true*, there's a duplicate birthday in the room. Now we can experiment with 1000 rooms and count the number of *true* results we see:

```
data:=[seq(evalb(nops({room(30)[]})<30),i=1..1000)]:
Statistics[Tally]( data );
```
$$[true = 709, false = 291]$$

That is, the probability that there are two people with the same birthday in a room of 30 randomly assembled people is about 709/1000 = 70.9% Are you surprised? Nearly everyone is surprised the first time they see this result. In fact, the theoretical probability that at least two people have the same birthday in a room of 30 people is approximately 70.6%. Our simulation was very close.

Other Distributions

The examples we have looked at so far assume a uniform distribution of integers over a range of integers. With the **Generate** command from the **RandomTools** package, Maple gives us considerably more flexibility in random number generation.

```
with( RandomTools ):
Generate( distribution( name( parameter ) ) );
```

We can use any distribution in the **Statistics** package.

Random Normal Variables

The normal distribution is used to describe things ranging from test scores to heights of adult lemmings. If we want to model a normally-distributed random event, we need to specify the mean and standard deviation. One of the popular standardized tests for American students is designed to have normally-distributed scores with mean 1500 and standard deviation 300. We can take a random sample, or make a list of 10000 random samples and see what the distribution looks like.

```
with(RandomTools):
Generate(distribution(Normal(1500, 300)));
```
$$2219.37395396710508$$

```
data:= Generate(list(
        distribution(Normal(1500, 300)), 10000) ):
Statistics[Histogram](data, frequencyscale=absolute);
```

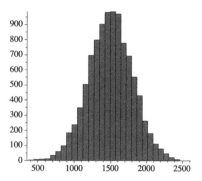

Random Binomial Variables

The binomial distribution is used to describe the number of successes in repeated random trials with a constant probability. The parameters for this distribution are the number of trials and the probability of success in each trial. For example we can look at a model where we have 10 trials with a 70% chance of succeeding in each trial. Once again we have Maple run 1000 trials and make a histogram of the results.

```
with(RandomTools): with(Statistics):
data:= Generate(list(
        distribution(Binomial(10, 0.7)), 1000)):
Histogram(data, frequencyscale=absolute);
```

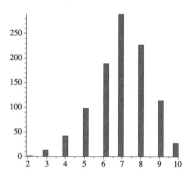

```
Tally(data);
```

$[8. = 227, 7. = 289, 2. = 1, 4. = 42, 6. = 189, 10. = 27, 3. = 13, 5. = 98, 9. = 114]$

Thus we see 7 is the most likely outcome in this model, and we got answers ranging from 2 to 10 in our 1000 trials.

Useful Tips

In this chapter, we presented only some basic methods with which you can generate random data from distributions having specified characteristics. You can find more tools for working with randomly generated objects in the **RandomTools** package.

The **Generate** command can be used to create random objects other than numbers. The following command provides a random 6th degree polynomial in x that has only positive, rational coefficients with denominators no larger than 12.

```
with( RandomTools ):
Generate( polynom( positive( denominator = 12),
                              x, degree = 6) );
```

$$\frac{11}{12} + \frac{5}{12}x + \frac{2}{3}x^2 + \frac{5}{6}x^3 + x^4 + \frac{11}{12}x^5 + \frac{1}{12}x^6$$

And, here's how to obtain a random bivariate quadratic polynomial with positive integer coefficients no larger than 20.

```
Generate(polynom(posint(range=20), {x,y}, degree=2));
```

$$3 + 6x + 3y + 14xy + 4x^2 + 3y^2$$

The online help for the **RandomTools** package, particularly the **Generate** command, provides extensive information (and examples) about using the **Generate** command.

Troubleshooting Q & A

Question... I cannot get the **Generate** command to work. What should I check?

Answer... Because **Generate** is defined in the **RandomTools** package, you need to make sure you used the **with(RandomTools);** command earlier in your session.

Question... How do I decide whether to use **rand** or **Generate** to get a random number?

Answer... The **Generate** command is generally preferred for its improved speed and flexibility. The **rand** command is simpler to use, but can only produce random integers. In fact, the implementation of the **rand** command involves calling routines in the **RandomTools** package.

Question... Why is there an extra set of parentheses at the end of nearly every occurrence of the **rand** command such as **rand(1..6)()** or **rand(10)()**?

Answer... The **rand(1..6)** or **rand(10)** command returns a function that, when called, produces a random number according to the arguments to the **rand** command. The extra parentheses tell Maple to execute the procedure produced by the **rand** command.

CHAPTER 27

Maple for Programmers

This chapter is designed for Maple users who are already familiar with writing computer programs. It is not a tutorial on programming. Rather, we expect that you are looking for information on those areas of Maple that are closely related to programming concepts.

Elements of Traditional Programming Languages

Maple has its own programming language that includes many of the programming elements found in traditional computer programs: conditional execution, looping, and subroutines. The syntax of the language borrows from the Pascal and C programming languages.

> **Note:** Until now, we have used the term "command" to describe Maple syntax (e.g., we used the **plot3d** command in Chapter 16). In this chapter, most commands will be called **statements**. This is more consistent with the terminology of programming languages.

The proc Statement

Almost every programming language supports a notion of a subroutine, function, or procedure. For example, BASIC allows the use of **DEF**, **GOSUB**, and **RETURN** statements, while FORTRAN has a **CALL** statement.

Subroutines are written in Maple using the **proc** statement. A common form is:

```
proc( parameter list )
    local local variable list;
    global global variable list;
    body
end proc;
```

The **local** and **global** clauses are optional. There are several additional optional clauses that can appear in procedure definitions. When the procedure is called, the statements in the body are executed in sequence.

The **return** statement can be used to specify the value returned by the procedure; in the absence of a **return** statement, the procedure returns the value of the last statement executed. A complete description of the general syntax can be found in Maple's help for the topic procedure (**?procedure**).

■ **Example.** To compute the area of a triangle with sides having length a, b, and c, you can use Heron's Formula which says that $area = \sqrt{s(s-a)(s-b)(s-c)}$, where $s = (a+b+c)/2$ is the semi-perimeter of the triangle. A nice coding of an area function (subroutine) would be:

```
area := proc( a, b, c )
  local s;
  s := (a+b+c)/2;
  sqrt( s*(s-a)*(s-b)*(s-c) );
end proc:

area(5, 5 ,8);
                 12
```

> **Note:** Don't forget to include semicolons or colons between statements in the body of a **proc**. Otherwise, Maple stops with a syntax error.

Since the length of each side should be positive, we can modify the parameter list in our **area** subroutine as follows to enforce the requirement that all three of the input parameters are positive.

```
area := proc(a::positive,b::positive,c::positive)
```

An error message is produced if one or more of the input variables is not positive.

```
area(-1,2,3);
Error, invalid input: area expects its 1st argument, a, to be
of type positive, but received -1
```

The print and printf Statements

Maple does not display the results of individual statements within a procedure. To have a procedure show more than its return value, we use a **print** statement:

```
print( expression1, expression2, ... );
```

Maple's **printf** statement, which is very similar to the C command of the same name, provides greater control of the appearance of the printed output.

```
printf( format string, expression1, expression2, ... );
```

To control the amount of information automatically displayed within a procedure, see the online help for **printlevel**.

■ **Example.** We can write a routine that computes the maximum, minimum, and average of a list of exam scores from a Calculus class, and then reports the scores in increasing order.

```
summarize := proc(data)
  local sortedList, n:
  sortedList := sort(data):
  n := nops(data);
  printf(`\nThe number of students is
%a.\nThe maximum score is %a; the minimum
score is %a.\nThe mean is %6.4f.`,
       n, sortedList[n], sortedList[1],
       add(data[k], k=1..n)/n );
  print( sortedList[] );
end proc:

class1 :=
[61,23,14,78,91,33,12,44,72,79,82,81]:
```

```
summarize( class1 );
```

The number of students is 12.

The maximum score is 91; the minimum score is 12.

The mean is 55.8333.

$$12, 14, 23, 33, 44, 61, 72, 78, 79, 81, 82, 91$$

In the format string for **printf**, **\n** generates a new line, **%a** is a placeholder for a generic Maple object, and **%f6.4** prints an object as a floating-point number with 4 decimal places and a maximum of 6 digits (not counting the decimal point or sign). See the online help for **printf** for descriptions of all the formatting specifications.

The if Statement

The most commonly used form of the **if** statement is:

> **if** *logical condition* **then** *result1*; **else** *result2*; **end if**;

If the *logical condition* evaluates as *true*, then *result1* is returned. If the *logical condition* evaluates as *false*, *result2* is returned. If Maple is unable to determine if the *logical condition* is *true* or *false*, the return value is *FAIL*. A full explanation of this three-valued logic can be found in Maple's help information for the topic boolean (**?boolean**). For example:

```
myFunction := proc(a)
  if a<=3 then 2*a-1 else 5*a+1 end if;
end proc:

myFunction( 3 );
                5
```

In an **if** statement, each of *result1* and *result2* can be either a single statement or a collection of statements separated by semicolons. The logical decision can also have more than two branches by using the optional **elif** clause (for "else if"):

> **if** *logical condition1*
>   **then** *result1*;
>   **elif** *logical condition2*
>   **then** *result2*;
>   **else** *result3*;
> **end if**;

Consider, for example, a procedure to solve the quadratic equation $ax^2 + bx + c = 0$:

```
roots2 := proc( a, b, c )
  local disc, r1, r2;
  disc := b^2-4*a*c;
  if disc>0 then
    r1 := (-b+sqrt(disc))/2/a;
    r2 := (-b-sqrt(disc))/2/a;
    printf(`The equation has two solutions at %a
           and %a.`, r1, r2 );
  elif disc=0 then
    r1 := -b/2/a;
    printf( `The equation has a double solution
           at %a.`,r1 );
  else
    printf(`The equation has no real solutions.`);
  end if;
end proc:
```

```
roots2( 4, 7, 2 );
```

The equation has two solutions at
           -7/8+1/8*17^(1/2) and -7/8-1/8*17^(1/2).

```
roots2( 4, 4, 1 );
```

The equation has a double solution at -1/2.

```
roots2(1, 1, 1 );
```

The equation has no real solutions.

The do Statement

The general form of the **do** statement is:

> **for** *index variable*
> **from** *expression*
> **to** *expression*
> **by** *expression*
> **while** *expression*
> **do**
>   *body*
> **end do:**

In this statement only the **do** *body* **end do:** is required; each of the other phrases is optional. If **from** or **by** is omitted, the default values of **from 1** and **by 1** are used.

The **to** expression and **while** expression are tested at the beginning of each loop. Both can be used at the same time. The **by** value is used to increase the index variable at the end of the loop.

The *body* can be either a single Maple statement or a sequence of statements separated by semicolons.

You can use the **next** statement to terminate the processing for the current index value and to move ahead to the next value of the index variable. To keep the loop from executing forever, you can use the **break** statement within the *body* of the **do** statement.

■ **Example.** The **nextprime** command can be used in a **do** loop to make a short list of primes:

```
aprime := 2;
do
  aprime := nextprime(aprime);
  if aprime > 12 then break end if;
end do;
```
                    2
                    3
                    5
                    7
                    11
                    13

The for Clause

Using a **do** loop with one or more **break** or **next** clauses to stop the iteration is rather inelegant in programming. The **for**, **from**, and **while** clauses provide more control over loops.

■ **Example.** The Fibonacci numbers are $c_0 = 1$, $c_1 = 1$, and $c_n = c_{n-1} + c_{n-2}$, for $n \geq 2$. A simple implementation of this definition is:

```
c := Array( 0..25 ):
c[0] := 1: c[1] := 1:
for i from 2 to 25 do
  c[i] := c[i-1] + c[i-2];
end do:
```

To see a few of the Fibonacci numbers, you can use:

```
seq( c[i], i=0..10 );
```

$$1, 1, 2, 3, 5, 8, 13, 21, 34, 55, 89$$

```
c[25];
```

$$121393$$

■ **Example.** The sum of the positive odd integers $1, 3, 5, \ldots, 49$, can be found with:

```
intsum := 0:
for i from 1 to 50 by 2 do
  intsum := intsum + i
end do:
intsum;
```

$$625$$

The while Clause

The same sum can be formed using the **while** clause as follows:

```
intsum := 0:
for i from 1 by 2 while i <= 50 do
  intsum := intsum + i;
end do:
intsum;
```

$$625$$

Note that this example is for illustration purposes only; to add a finite sequence of consecutive odd integers it would be better to use:

```
add( 2*i-1, i=1..25 );
```

$$625$$

Recursive Structures in Maple

A recursive procedure is one that calls itself. In a recursive procedure it is necessary to ensure that the recursion terminates after a finite number of steps. Let's now explore some simple uses of recursion within Maple.

■ **Example.** The Fibonacci numbers are defined recursively as $c_0 = 1$, $c_1 = 1$, and $c_n = c_{n-1} + c_{n-2}$, for $n \geq 2$. A simple recursive Maple implementation of the Fibonacci numbers is:

```
c := proc( n::nonnegint )
   if n < 2 then 1
             else c(n-1) + c(n-2)
   end if
end proc:
```

The procedure call checks that **n** is a nonnegative integer. The statement **if n < 2 then 1** will define **c(0) := 1** and **c(1) := 1**. To find the Fibonacci number c_5, just type:

```
c(5);
```
$$8$$

To evaluate **c(5)**, Maple repeatedly uses the three rules above to simplify the expression in essentially the following sequence of transitions:

$$c_5 \rightarrow c_4 + c_3 \rightarrow (c_3 + c_2) + (c_2 + c_1) \rightarrow ((c_2 + c_1) + (c_1 + c_0)) + ((c_1 + c_0) + 1)$$

$$\rightarrow (((c_1 + c_0) + 1) + (1 + 1)) + ((1 + 1) + 1) \rightarrow (((1 + 1) + 1) + 2) + (3) \rightarrow 8$$

Remembering Values During Recursion

One disadvantage of the method above is that Maple will not remember the values that it has calculated recursively. For example, if you want to calculate **c(6)**, Maple recomputes **c(5)**, **c(4)**, **c(3)**, and **c(2)** all over again. This can slow down computations dramatically even for, say, **c(20).**

The **remember** option of Maple gives you a way to evaluate **c(5)** and at the same time define its value in case you need to use it later. You do this with the following new definition of the Fibonacci procedure:

```
newFib := proc( n::nonnegint )
   option remember:
   if n < 2 then 1
             else newFib(n-1) + newFib(n-2)
   end if
end proc:
```

This option makes Maple create a "remember table" so that **newFib(i)** is only computed once for each input value of **i**. On subsequent calls, **newFib(i)** is evaluated with a "lookup table."

Using option **remember** can have a dramatic effect on execution time. It takes about the same amount of time to complete the function calls **c(25)** and **newFib(20000)**.

Viewing the Code of Maple Procedures

All but the most basic Maple statements are written in the Maple programming language. Almost all of the code in Maple is readily available to users. This means you can see exactly what Maple does when you execute a command. It also makes it relatively easy to modify or extend a command for some specific purpose.

The showstat Command

The easiest way to view Maple code is with the **showstat** (<u>show stat</u>ements) command.

■ **Example.** To see the code for the Maple command **norm**:

```
showstat( norm );

norm := proc(p::{Matrix, Vector, matrix, polynom, vector},
n::satisfies(t -> evalb(1 <= t)), v)
local c;
   1   if type(p,'array') then
   2      return linalg['norm'](args)
       elif type(p,'rtable') then
   3      return LinearAlgebra:-Norm(args)
       end if;
   4   if nargs < 2 .......
```
(output truncated)

Another way of showing code in a Maple procedure is to change the value of the system variable **verboseproc** with the **interface** command. This variable is usually set at **1**. To see code in a procedure the variable should be set to **2**. Setting the variable to **3** will let you see remember tables as well. Then, use the **print** command to see the code. For example to see the code for **newFib** that we defined above.

```
interface( verboseproc=3 );
print( newFib );
```
(output not shown)

To understand what a block of code does, it is useful to trace through values of the assigned variables as the procedure executes. The option to trace a procedure is turned on with the **debug** command and turned off with the **undebug** command.

```
debug( newFib ):
newFib( 10 );
undebug( newFib );
```
(output not shown.)

You should be warned that most common commands are composed of several pages of code, as Maple carefully breaks most commands into many cases.

File I/O

Saving Values or Procedures for Future Maple Use

You can **save** results from one session of Maple and then **read** them back in during a future Maple session. The statement

```
save( newFib, "mondayMaple.m" );
```

saves the procedure **newFib** in a binary file named **mondayMaple.m**. The file would be read back in during a later Maple session with

```
restart;
read( "mondayMaple.m" );
```

This is an internally formatted Maple file and is not readable with a text editor. It is, however, in ASCII so that it can be transferred via e-mail.

The **currentdir** command is used to check, and change, the current directory where the **save** and **read** statements look for their files. The initial directory is the full path to the folder from which Maple was launched.

Importing Data from a File

To import a data file to Maple, you will use the **readdata** command. Suppose you have a text file called **file1.txt** with the data one to a line. You can type:

```
readdata( "file1.txt" );
```

to import the data. The **readdata** statement accepts arguments to specify the exact format of the input.

Exporting Data to a File

To export numeric data from Maple, you must go through a three-step process of opening a file, writing to it, and closing the file.

- First, you must create the file descriptor for the output file. For a buffered file, you use the **fopen** command in the form:

fopen(*file name*, READ/WRITE/APPEND, TEXT/BINARY **);**

For example, if we want to append text to **file2**, we would use the statement:

```
fd1 := fopen( "file2", APPEND, TEXT );
```

- Next, you use the **writedata** statement to write values to the file. You may write a list or vector or matrix. To write **myData** to **file2**, use the statement:

```
myData := [20, 31, 15, 26, 17, 18];
writedata( fd1, myData );
```

- Finally, when you are finished with output, you should **fclose** the file:

```
fclose( fd1 );
```

If you use a text editor to open **file2**, you will see the data 20, 31, etc., one to a line. You can read this file back to Maple using the **readdata** command.

```
readdata( "file2" );
```

$$[20, 31, 15, 26, 17, 18]$$

Optionally, you can add formatting information after the data. More than one **writedata** statement can be used between the **fopen** and **fclose**.

■ **Example.** Suppose that you want to compute the values of the Fibonacci numbers with the procedure **newFib** defined above, and then save those values to a file named **"Fibonacci.dat"**, first with all 25 values on one line, then with 25 lines each containing the list **[i, newFib(i)]**.

```
fd1 := fopen( "Fibonacci.dat", WRITE, TEXT ):
writedata( fd1, [[seq(newFib(i),i=1..25)]] );
for i from 1 to 25 do
    writedata( fd1, [[i,newFib(i)]] );
end do;
fclose( fd1 );
```

To read the data we can use **readdata**.

```
readdata( "Fibonacci.dat" );
```

Translating Maple into Other Programming Languages

C, Fortran, Java, MATLAB, and VisualBasic

After we have worked with a procedure in Maple, we may want to use the code to write a program in another computer language. This can be done with commands from the **CodeGeneration** package. The Maple commands **Matlab**, **Fortran**, **Java**, **C**, and **VisualBasic**, produce the corresponding code equivalents of Maple expressions and procedures.

■ **Example.** Earlier in this chapter we saw a procedure for finding the area of a triangle using Heron's formula. Here are the automatic translations into C and MATLAB.

```
area := proc( a::positive, b::positive, c::positive )
  local s;
  s := (a+b+c)/2;
  sqrt( s*(s-a)*(s-b)*(s-c) );
end proc:
```

```
with( CodeGeneration ):
C( area );
#include <math.h>
double area (double a, double b, double c)
{
  double s;
  s = a / 0.2e1 + b / 0.2e1 + c / 0.2e1;
  return(sqrt(s * (s - a) * (s - b) * (s - c)));
}
```

```
Matlab( area );
function areareturn = area(a, b, c)
  s = a / 0.2e1 + b / 0.2e1 + c / 0.2e1;
  areareturn = sqrt(s * (s - a) * (s - b) * (s - c));
```

Translations into Fortran, Java, and VisualBasic are obtained in the same manner. Each of these commands has options to optimize the code and to save the results directly to a file.

Notice that neither of these translations includes a check that the arguments are positive. This is an example of a Maple programming element that cannot be automatically translated into another language. Sometimes it is necessary to modify the Maple code to handle this explicitly.

■ **Example.** The procedure **newFib** above for finding Fibonacci numbers uses a remember table for increased efficiency. This structure does not translate into other languages. The **CodeGeneration** package produced code with a comment that indicates that "option" is not translated.

```
Java( newFib );
Warning, the function names {newFib} are not recognized in the
target language
Warning, procedure/module options ignored
```

We can modify the statement.

```
newFib2 := proc( n )
if n<0 then
  error("invalid input, non-negative integer argument
needed")
end if;
  if n < 2 then 1
           else newFib2(n-1) + newFib2(n-2)
  end if
end proc:
```

```
Java( newFib2 );
```

```
Warning, the function names {newFib2} are not recognized in the
target language
class CodeGenerationClass {
  public static int newFib2 (int n)
  {
    if (n < 0)
      throw new Exception("invalid input, non-negative integer
argument needed");
    if (n < 2)
      return(1);
    else
      return((int) (newFib2(n - 1) + newFib2(n - 2)));
  }
}
```

The warning about an unrecognized name in the target language is produced because this procedure is recursive.

Maple also allows you to make external calls to compiled Fortran, C, or Java code, but that feature is beyond the scope of this book. For more information on this, search for **define_external** in the Maple **help** menu.

APPENDIX A
Learning Calculus with Maple

Maple Resources on WWW

Maple can be an effective tool in learning mathematics, particularly calculus. There are many publicly-available educational Maple resources, for both students and instructors. The largest collection of web-based resources is found in the Maple Application Center:

> http://www.mapleapps.com/

MaplePrimes is a web-based community dedicated to sharing experiences, techniques, and opinions about Maple. To join this community, visit

> http://www.mapleprimes.com/

The Student[Calculus1] package

The **Student[Calculus1]** package contains many commands helpful in learning Calculus. Some of these commands provide animations or graphs that demonstrate Calculus concept. Others provide interactive Maplets to help students practice applying the limit, differentiation, and integration rules step-by-step.

Visualization Resources

Here we will give a quick summary on some of those visualization resources inside the **Student[Calculus1]** package. You should first load the package:

> **with(Student[Calculus1]):**

Visualization Commands for Learning About Derivatives

Example	Comment
`Tangent( sin(x), x=1, output=line);` `Tangent( sin(x), x=1, output=plot);`	**output=line** will give the equation of tangent line at the given point while **output=plot** will draw the graph of both the function and tangent line.
`increment := [0.5, 0.1, 0.01, 0.001];` `NewtonQuotient( sin(x), x=1,` `  output=value, h=increment);` `NewtonQuotient( sin(x), x=1,` `  output=animation, h=increment);`	**output=value** will compute the values of Newton Quotient at the given increment, while **output=animation** will illustrate graphically how the secant lines at the given increment points converge to the tangent line.
`RollesTheorem( sin(x), x=1..1+2*Pi,` `  output=points);` `MeanValueTheorem( sin(x), x=0..3,` `  output=plot);`	**output=points** will compute the value(s) of the point(s) that satisfies the specified theorem at the given interval, while **output=plot** will illustrate graphically the specified theorem at the given interval.

Visualization Commands for Learning About Applications of the Derivative

Example	*Comment*
`DerivativePlot( (x-5)*exp(x), ` `  x=-3..3, order=1..5);`	The given example will sketch the graphs of the given function and its first 5 derivatives in a single picture for *x* between –3 and 3.
`FunctionChart( x^3-9*x^2-48*x+52, ` `  x=-6..14);`	It will indicate in the picture, the regions where the function is increasing/decreasing, and concave up/down. Also it will indicate the roots, local max/min points and inflection points of the given function.

`TaylorApproximation( sin(x), x=1, `
`  order=3);`
 	
`TaylorApproximation( sin(x), x=1, `
`  output=animation, order=1..10);`
	The first example will give the Taylor approximation of order 3 at the point *x*=1.
	
	With **output=animation, order=1..10**, we can see how the approximation is improved as the order is increased from 1 to 10.

`NewtonsMethod( sin(x), x=2, `
`  output=sequence);`
 	
`NewtonsMethod( sin(x), x =2, `
`  output=plot);`
	output=sequence will generate a sequence of points from the Newton's method to approximate the zero of a function starting at the given point, while **output=plot** will illustrate graphically how these points are selected from the Newton's method.

Visualization Commands for Learning About Integrals

Example	*Comment*

`AntiderivativePlot( x^3 +2*x^2 −1, `
`  x=-2..2);`
 	
`AntiderivativePlot( x^3 +2*x^2 −1, `
`  x=-2..2, showclass);`
	The first command will draw the given function together with one of its antiderivatives. When we add the **showclass** option in the second command, Maple will draw a family of antiderivatives.

`ApproximateInt( x^2, x=0..4, ` `method=midpoint, output= plot);`	It will sketch the graph of the given function (in the specified interval) together with the rectangles that approximated the definite integral. Options for **method** include **left**, **right**, **trapezoid**, **lower**, **upper**, **random**, **simpson** and **newtoncotes**. If we use **output=animation** instead, then the picture will demonstrate that as the number of rectangles increased, the definite integral is better approximated.

Visualization Commands for Learning About Applications of Integration

Example	*Comment*
`FunctionAverage(sin(x), x=0..Pi, ` `  output=plot);`	It will sketch the graph of the given function and its average value in the specified interval.

`VolumeOfRevolution( sin(x)+2, `
`  x=0..Pi, output=plot);`
 	
`SurfaceOfRevolution( sin(x)+2, `
`  x=0..Pi, output=plot);`
	These commands will draw the object corresponding to the volume or surface of revolution of the given function along the horizontal axis. If we want to rotate the function along the vertical axis, we need to include the option, **axis=vertical**.

LimitTutor Commands

The **Student[Precalculus]** and **Student[Calculus1]** packages each has its own **LimitTutor** command that helps the students understand limits. For example to understand $\lim_{x \to 0} \dfrac{\sin(\sin(2x)^2)}{x^2}$, we can enter:

```
with(Student[Precalculus]):        #Precalculus version
LimitTutor(sin(sin(2*x)^2)/x^2, x=0);
```

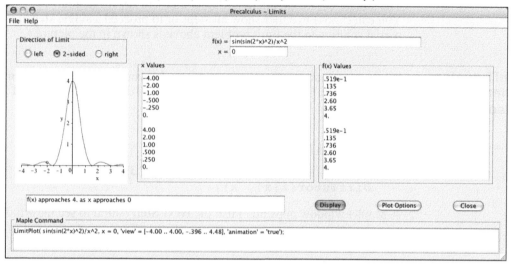

```
with(Student[Calculus1]):          #Calculus version
LimitTutor(sin(sin(2*x)^2)/x^2, x=0);
```

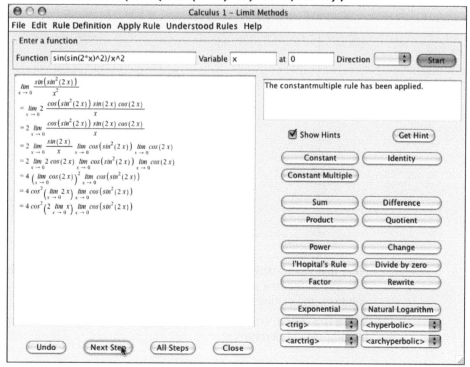

The **Precalculus** version of **LimitTutor** approaches limits in an intuitive manner. The command calls up a Maplet that shows the graph of the function, along with the value of the function at a set of points approaching the place where we want to take the limit.

The **LimitTutor** command inside **Calculus1** takes a more analytical approach. The similar command opens a Maplet that will apply theorems to evaluate the limit. You can choose to either show all steps or to proceed a single step at a time. A partial solution to the limit problem above is shown in the picture.

DiffTutor for intermediate steps

When differentiating, it is important to understand the various differentiation rules. The **Student[Calculus1]** package lets us do that with the **DiffTutor** command.

For example,

```
with(Student[Calculus1]):
f := x -> x*sin(3*x+x^2)/(x^3+x);
DiffTutor(f(x), x);
```

A Maplet window pops up. Push the **Next Step** button to see the window below.

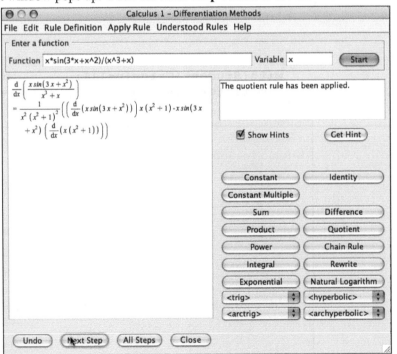

Note that we needed to start with the quotient rule, and we see the result of using that rule as the first step. Repeated use of the **Next Step** button gives a complete solution, and stating which rules to use during the process.

There is a similar command called **IntTutor** that provides step-by-step procedures to compute antiderivatives.

APPENDIX B

Quick Reference Guide to Maple 12

Mathematical Operations, Constants, and Functions:

Symbol	Description	Example
+, -, *, /, ^, !	add, subtract, multiply, divide, power, factorial	`1-2*x^2+3*x^4/4!`
Pi, I, infinity	mathematical constants (π, i, ∞) Note: Maple is case sensitive	`exp( Pi*I );` `sum(1/k,k=1..infinity);`
sin, cos, tan, cot, sec, csc	trigonometric functions	`sin(x-Pi/5)-sec(x^2);`
arcsin, arccos, arctan, arccot, arcsec, arccsc	inverse trigonometric functions	`arctan( 2*x );`
exp	exponential function	`exp( 2*x );`
ln, log, log10, log[b]	logarithm functions (log is the natural log)	`ln( x*y/2 );` `log10( 1000 );`
abs	absolute value	`abs( (-3)^5 );`
sqrt, root[n]	(principle) square root, n^{th} root	`sqrt( 24 );`
=, <>, <, <=, >, >=	relational operators for equations and inequalities	`diff(y(x),x)=y(x);` `is(exp(Pi)>Pi^exp(1));`

Symbols and Abbreviations:

Symbol	Description	Example
:=	assignment	`f := x^2/y^3;`
;	command terminator with results displayed	`int( x^2, x );`
:	command terminator with results hidden	`int( x^2, x ):`
..	specify a range or interval	`plot(t*exp(-2*t),t=0..3);`
{ }	set delimiter	`{ y, x, y };`
[ ]	list delimiter	`[ y, x, y ];`
%	ditto operator	`Int( exp(x^2), x=0..1 ):` `% = evalf( % );`
" "	string delimiter (double quote, see ?strings)	`plot(sin(10*x)+x,x=0..Pi,` `    title="A neat plot");`
` `	name delimiter (back quote, see ?names)	`` `my func` := sin(x); `` `` int( `my func`, x ); ``
' '	delayed evaluation (single quote, see ?quotes)	`unassign( 'x' );`
->	mapping (procedure) definition	`f:= (x,y) -> x^2*sin(x-y);` `f(Pi/2,0);`

Some Frequently Used Commands:

Command	Description	Example
`restart`	clear all Maple definitions	`restart;`
`with`	load a Maple package	`with(plots): with(DEtools);`
`Help` (or `?`)	display on-line help	`?limit`
`limit`	evaluate a limit	`limit( sin(a*x)/x, x=0 );`
`diff`	differentiate an expression	`diff(a*x*exp(b*x^2)+cos(c*y),x);`
`int`	definite or indefinite integral	`int( sqrt(x), x=0..Pi );`

`Limit,`
`Diff,`
`Int`
	inert (unevaluated) forms of `limit`, `diff`, and `int`	`Limit( sin(a*x)/x, x=0 );`
		`Diff(a*x*exp(b*x^2)+cos(c*y),x);`
		`q := Int( sqrt(x), x=0..Pi ):`

`value`	evaluate an inert expression	`q = value(q);`
`D`	differentiation operator	`D(cos);`
`plot`	create a 2-dimensional plot	`plot(u^3,u=0..1,title="cubic");`
`plot3d`	create a 3-dimensional plot	`plot3d(sin(x+y),x=0..4*Pi,y=0..Pi);`

`display`	combine multiple plot structures (**plots** package)	`F:=plot(exp(x),x=-3..3):`
		`G:=plot(1+x+x^2/2,x=-3..2):`
		`plots[display]([F,G]);`

`solve`	solve equations or inequalities	`solve( x^4-5*x^4+6*x=2, x );`
`dsolve`	solve ordinary differential equations	`dsolve({diff(y(x),x,x)-y(x)=1,` `        y(0)=A,D(y)(0)=B}, y(x));`

`odeplot`	create plots from numerical solution to an IVP found with **dsolve**	`S :=diff(x(t),t)=-y(t),`
		`       diff(y(t),t)=x(t):`
		`IC:=x(0)=1, y(0)=1:`
		`P:=dsolve({S,IC},{x(t),y(t)},numeric):`
		`plots[odeplot](P,[x(t),y(t)],0..Pi);`

`DEplot`	create plot for ODEs (**DEtools** package)	`ODE:=diff(y(x),x)=2*x*y(x);` `DEplot(ODE, y(x), x=-2..2, y=-1..1);`
`simplify`	apply simplification rules to an expression	`simplify( exp(a+ln(b*exp(c))) );`
`factor`	factor polynomial	`factor((x^3-y^3)/(x-y));`
`convert`	convert expression to a different form	`convert( x^3/(x^2-1), parfrac, x);`
`collect`	collect coefficients of like powers	`collect((x+1)^3/(x+2),x);`
`rhs,` `lhs`	extract the right- and left-hand sides of an equation	`rhs( y=a*x^2+b*x+c );` `lhs( y=a*x^2+b*x+c );`
`numer,` `denom`	extract the numerator and denominator of a rational expression	`numer( (x+1)^3/(x+2) );` `denom( (x+1)^3/(x+2) );`
`eval`	evaluate an expression with specific values	`eval(sin(x)/cos(x),x=0);`
`evalf`	floating-point evaluation of an expression	`evalf( exp(Pi^2) );`
`evalb`	evaluate a Boolean expression (result is **true**, **false**, or **FAIL**)	`evalb( 2^8>200 );`
`evalc`	evaluate a complex-valued expression (result has form **a+I*b**)	`evalc( exp(x+I*y) );`
`seq`	create a sequence	`seq([0,i],i=-3..3);`

`assign`
`unassign`
	make and remove assignments	`solve({x+y=1,x-y=3},{x,y});`
		`assign( % ); x, y;`
		`unassign('x','y'); x, y;`

Index

Printed and bound by CPI Group (UK) Ltd, Croydon, CR0 4YY

20/10/2024

14576725-0002